Fires in GunaiKurnai Country

Fires in GunaiKurnai Country

Landscape Fires and their Impacts on Aboriginal Cultural Heritage Places and Artefacts in Southeastern Australia

Edited by

Jessie Buettel, Bruno David, Russell Mullett,
Joanna Fresløv, Katherine Szabó,
GunaiKurnai Land and Waters Aboriginal Corporation

ARCHAEOPRESS ARCHAEOLOGY

ARCHAEOPRESS PUBLISHING LTD
Summertown Pavilion
18-24 Middle Way
Summertown
Oxford OX2 7LG
www.archaeopress.com

ISBN 978-1-80327-481-2
ISBN 978-1-80327-482-9 (e-Pdf)

Cover: On Country fire occurs with the right fire, at the right time and in the right way, and in the right place (taken for the Gunaikurnai Cultural Fire Strategy). Photograph by Jessica Shapiro, courtesy of GunaiKurnai Land and Waters Aboriginal Corporation.

Contents

List of Figures and Tables

Introduction

Wildfires: Characteristics, Drivers and Impacts on Cultural Sites

Accounts and Memories of Landscape Burning Practices in Gippsland

Eugene von Guérard on GunaiKurnai Country 1860–1861: Reading the Story of Fire in his Depictions of the Landscape

20th and 21st Century Wildfires and Prescribed Burning in GunaiKurnai Country

Cultural Sites in GunaiKurnai Country

The Impacts of Fire on Stone Artefacts

The Impacts of Fires on Rock Art Sites and Ochre

The Impact of Fires on Bone

The Impacts of Fire on Culturally Modified Trees

Shells and Fire—Indicators and Effects

Landscape Fires and Cultural Sites in GunaiKurnai Country

Archaeological Surveys in GunaiKurnai Country

Understanding the Distribution and Impacts of Wildfires in GunaiKurnai Country through Subregions

Authors and Affiliations

Jessie Birkett-Rees: Centre for Ancient Cultures, Monash University, Australia.

Jessie Buettel: School of Natural Sciences, University of Tasmania, Australia; Australian Research Council Centre of Excellence for Australian Biodiversity and Heritage, Australia.

Bruno David: Monash Indigenous Studies Centre, Monash University, Australia; Australian Research Council Centre of Excellence for Australian Biodiversity and Heritage, Australia.

Jean-Jacques Delannoy: EDYTEM, Université Savoie Mont Blanc, France; Australian Research Council Centre of Excellence for Australian Biodiversity and Heritage, Australia.

Joanna Fresløv: GunaiKurnai Land and Waters Aboriginal Corporation, Australia.

GunaiKurnai Land and Waters Aboriginal Corporation: Kalimna West, Australia.

Jillian Huntley: Griffith Centre for Social and Cultural Research, Griffith University, Australia.

Matthew McDowell: Monash Indigenous Studies Centre, Monash University, Australia; Australian Research Council Centre of Excellence for Australian Biodiversity and Heritage, Australia.

Jerome Mialanes: Monash Indigenous Studies Centre, Monash University, Australia; Australian Research Council Centre of Excellence for Australian Biodiversity and Heritage, Australia.

Russell Mullett: GunaiKurnai Land and Waters Aboriginal Corporation, Australia.

Annette Oertle: Department of Evolutionary Anthropology, University of Vienna, Austria.

Stefania Ondei: School of Natural Sciences, University of Tasmania, Australia.

Ruth Pullin: School of Culture and Communication, The University of Melbourne, Australia.

Robert Skelly: Monash Indigenous Studies Centre, Monash University, Australia.

Seumas Spark: School of Philosophical, Historical and International Studies, Monash University, Australia.

Katherine Szabó: Pre-Construct Archaeology, Cambridge, United Kingdom.

Courtney Webster: School of Languages, Humanities and Social Science, Griffith University, Australia.

Grant Williamson: School of Natural Sciences, University of Tasmania, Australia.

Acknowledgements

The research presented in this volume was funded by a grant from the Department of Land, Water and Planning (Victorian State government) and Parks Victoria to the GunaiKurnai Land and Waters Aboriginal Corporation, through the Bushfire Biodiversity Response and Recovery (BBRER) program. We are grateful for their support in this.

Jessie Buettel and Bruno David thank the Australian Research Council Centre of Excellence for Australian Biodiversity and Heritage project CE170100015 (Monash University and University of Tasmania Nodes) for support.

Seumas Spark thanks Bill Gammage for providing a copy of Francis Peter MacCabe's map of 1847 and pointing out that it represents a fire-managed landscape. He further thanks Chris Commins for providing a copy of the report by The Mountain Cattleman's Association of Victoria, and Vic Jurskis for information on the Thomas Townsend quote cited in Chapter 3.

Ruth Pullin thanks Professor Bruno David for his editorial suggestions, generous advice and contributions to Chapter 4.

Jerome Mialanes thanks Geraldine Fiers, Fernanda Neubauer and Patrick Schmidt for their assistance in providing figures for Chapter 7.

Kat Szabó and Annette Oertle thank Cygnet Bay Pearl Farm, Broome, Western Australia, for the generous provision of pearl oysters used in the experimental work presented in Chapter 11. Szabó's research was funded by an Australian Research Council Future Fellowship (FT140100504) and Oertle's research was supported by an Australian Postgraduate Award.

The authors would like to deeply thank the GunaiKurnai Land and Waters Aboriginal Corporation for the invitation, privilege and opportunity to work together on this volume. Thank you for showing and inviting us to learn about your Country.

Chapter 1

Introduction

Bruno David, Russell Mullett, Joanna Fresløv
and the GunaiKurnai Land and Waters Aboriginal Corporation

Wildfires, often called 'bushfires' in Australia, are an annual occurrence in Australia, especially in the southeast. Each year they burn thousands of square kilometres of bushland, blazing everything in their way. Some of the physical infrastructure erected or grown by people, such as fencelines, buildings and crops can be rebuilt and replanted, despite the devastation wildfires leave behind for the communities whose lives are upended by their damage or destruction. But other things cannot. Among these latter are the Aboriginal cultural heritage places that have accumulated in the landscape through the course of history and that often form the only material expressions of thousands of years of culture. How do these wildfires impact on Aboriginal sites? Do some site types, or materials, fare better than others? And how have wildfires contributed to a diminution or erasure of material culture, such as those that would otherwise be seen in the landscape as one went about in everyday life, and that contributes an essential, often subliminal role in cultural education down the generations?

With the above concerns in mind, this monograph examines the incidence of landscape fires in relation to the distribution of archaeological sites across one part of southeastern Australia, the lands of the GunaiKurnai Aboriginal clans. In doing so, we review what is known of the region's fire history through a number of sources: through oral histories, colonial art, and through an examination of government fire records dating back to the early years of the 20th century. The stimulus for this work were particularly severe wildfires that raged across much of southeastern Australia, including GunaiKurnai Country, in the spring and summer of 2019–2020. Our aim is to better understand the impacts of wildfires across GunaiKurnai Country's varied environments, so as to be in a better position to find some answers for the management of its cultural heritage sites and landscapes.

THE GIPPSLAND FIRES OF 2019–2020

November 2019 was a particularly hot and dry month across many parts of southern Australia. Over a four-day period, 18–21 November, temperatures reached record highs. On 21 November, 150 wildfires covering 326,000 hectares were recorded across Victoria; 60 of these fires were still burning by the end of the day, three of the largest in East Gippsland, much of which lies in GunaiKurnai Country. The Bruthen and Gelantipy fires in particular continued to grow over the coming days.

A month later, on 20 December, a new heatwave led to 110 additional wildfires, again prominently in East Gippsland, with an area of 15,000km^2 being particularly at risk. By 30 December, a host of new fires had burst through. 'Three fires in [East Gippsland] with a combined area of more than 130,000ha remained active; some fires burned with sufficient intensity to

Figure 1.1. Wildfire front burning in eucalypt forest near Buchan, East Gippsland, 2019–2020. Photograph my Shamis Law, courtesy of Matt Holland.

create pyrocumulonimbus clouds that generated local thunder and lightning' (AIDR 2020). The fires continued to burn over the coming days and weeks, so that by 27 February 2020, as the fires waned, 1.5 million hectares of land had burned. The calamitous loss of approximately 120 lives through smoke inhalation, over 1000 cases of hospitalisation, and disastrous damage to property, resources, wildlife and habitats had reached unprecedented levels (for a summary of impacts from the 2019–2020 Victorian wildfires, see AIDR 2020) (Figure 1.1).

While the 2019–2020 wildfires were particularly severe and widespread, GunaiKurnai Country has suffered a number of major wildfires over the past century. The Black Friday fires of 1939 burnt 35% of GunaiKurnai Land and Waters Aboriginal Corporation (GKLaWAC) lands; the Gippsland fires of 1965, 17%; the Great Divide fires of 2007, 27%; the Gippsland fires of 2019–2020, 20%. These individual fires (wildfires in rapid succession) all burnt huge areas of land. They had severe impacts at the time, and have had continued or cumulative effects on people's lives. But what has not yet been documented are the compounding impacts of the 2019–2020 and earlier wildfires—their frequency, location and extent—on Aboriginal cultural sites in GunaiKurnai Country.

In light of these calamitous wildfires, and under the already-proven success of an established Memorandum of Understanding for partnership research between GKLaWAC and Monash University, a few months after the 2019–2020 wildfires GKLaWAC commissioned the authors to undertake a desktop study of the distribution of registered cultural sites across the geographical spread of the 2019–2020 Gippsland and earlier wildfires in the GKLaWAC Registered Aboriginal Party (RAP) area (a RAP is a legally recognised representative Aboriginal group with formal rights to protect and manage Aboriginal cultural heritage on their lands and waters). At Monash University, this work was done through the Monash Indigenous

Studies Centre and the Australian Research Council Centre of Excellence for Australian Biodiversity and Heritage. The study was undertaken in partnership with GKLaWAC as part of the Victorian government's Bushfire Biodiversity Response and Recovery (BBRER) program, and as a background step in community-led on-Country cultural heritage and biodiversity wildfire response and recovery. The study's major aims were to understand the impacts of landscape fires on cultural sites across GunaiKurnai Country. This monograph is one outcome of this study, and brings together a broad range of information on fires and their impacts on cultural sites and cultural materials. While it is designed as a one-stop resource for GKLaWAC and cultural heritage managers in GunaiKurnai Country and Australia more generally, some results (e.g., on the effects of fires on shell) and methodologies (Geographical Information System (GIS) analyses of wildfires and their effects on sites) have broader, international significance or adaptability.

THE GUNAIKURNAI LAND AND WATERS ABORIGINAL CORPORATION: PEOPLE AND COUNTRY

The study area covers the whole of the GKLaWAC RAP area, being the Registered Aboriginal Party that represents the GunaiKurnai Aboriginal Traditional Owners of eastern Victoria, southeastern Australia. Comprising some 25,770km^2 (calculated using VicGrid94 projections), the GKLaWAC RAP area extends from the mountains of the Great Dividing Range at Mount Hotham (1861m above sea level) in the north to Bass Strait in the south (Figure 1.2). This is a diverse landscape, with only 125km between the High Country and the coastal lakes. In pre-colonial times prior to the early 1800s, GunaiKurnai Country was partly insulated from neighbouring groups by the mountain ranges to the north, the sea to the south, and dense temperate wet forests to the east. Many traditions and cultural practices of the GunaiKurnai clans differed from those of outside groups, who were referred to by GunaiKurnai as *brajerak*, 'aliens' or not-GunaiKurnai (Howitt 1904: 41).

The GunaiKurnai were traditionally divided into five dialect groups or 'clans': the Brayakaulung, Brataualung, and Tatungalung to the west; Brabralung in the central area; and Krauatungalung to the east, with smaller, fluid residential groups (sometimes called 'bands' in the literature) within each of these larger groups (Fison and Howitt 1880: sketch map; Howitt 1904: 73) (Figure 1.3). The five main clans each have Country that includes parts of the coast or the large lakes system, but only Brayakaulung, Brabralung and the Krauatungalung have mountain areas in their territories. The rugged nature of the High Country to the north meant that there were only a few travel routes through the mountains (Howitt in Smyth 1878, vol. II: 325). In Krauatungalung Country, where archaeological research has been focused since 2018 and where the 2019–2020 wildfires were particularly severe, the Snowy River (Doorack) was a major travel route between the coast and the mountains (Bulmer in Smyth 1878, vol. II: 191; Howitt 1904: 518, 693).

Understanding GunaiKurnai social relationships is key to understanding the archaeological sites and artefacts in the landscape, because people and goods travelled at least as much to maintain social connections and affiliations with different parts of Country as for food and other material resources. The activities, contents and locations of camps and other frequented places, now evident as archaeological and oral history sites and artefacts, were a product of those connections and patterns of mobility.

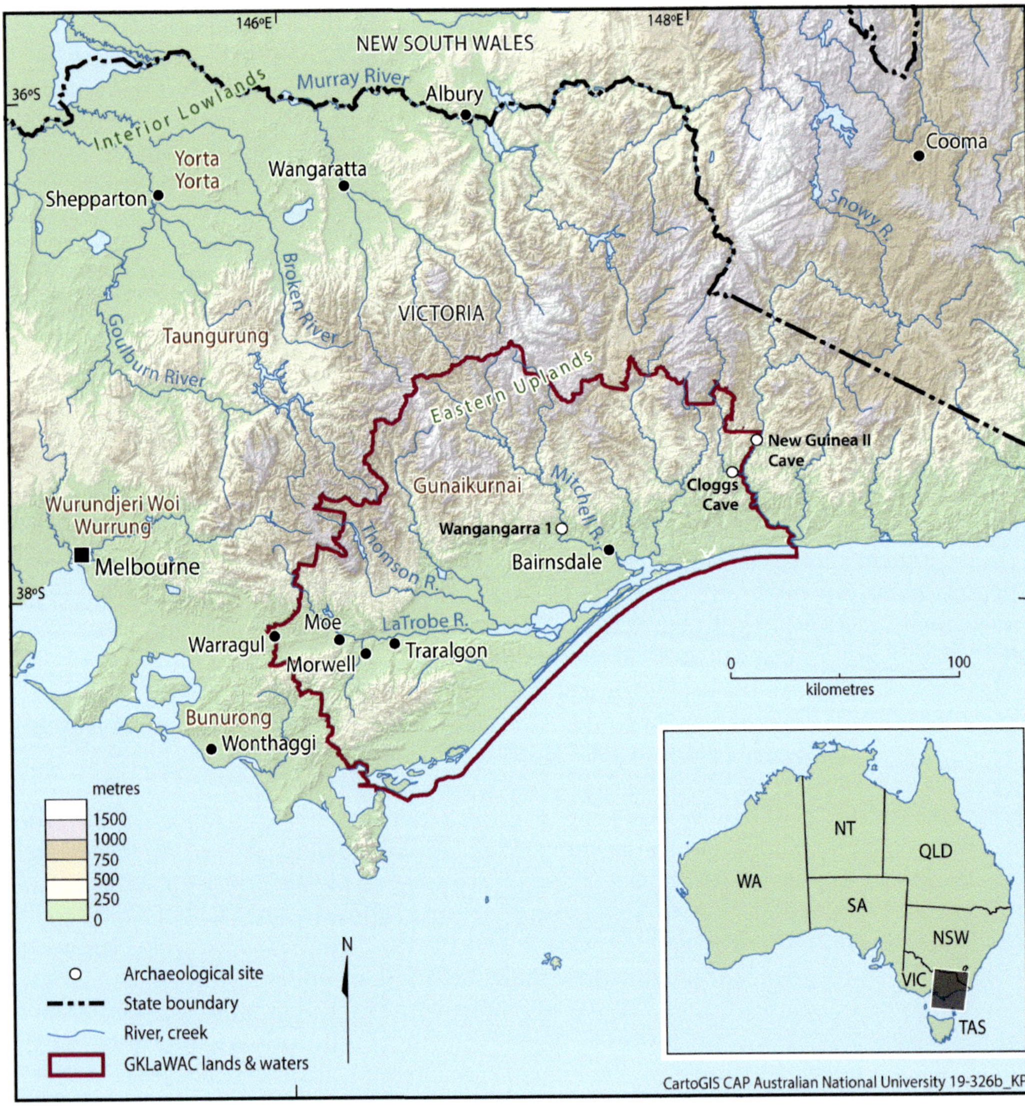

Figure 1.2. Location of the GKLaWAC RAP area, showing excavated archaeological sites where major findings have been made (figure by CartoGIS Services, College of Asia and the Pacific at the Australian National University, using Esri ArcMap 10.5 (https://desktop.arcgis.com/en/arcmap/) and Adobe Illustrator CC 2017 (21.0) https://helpx.adobe.com/au/illustrator/release-note/illustrator-cc-2017-21-0-release-notes.html).

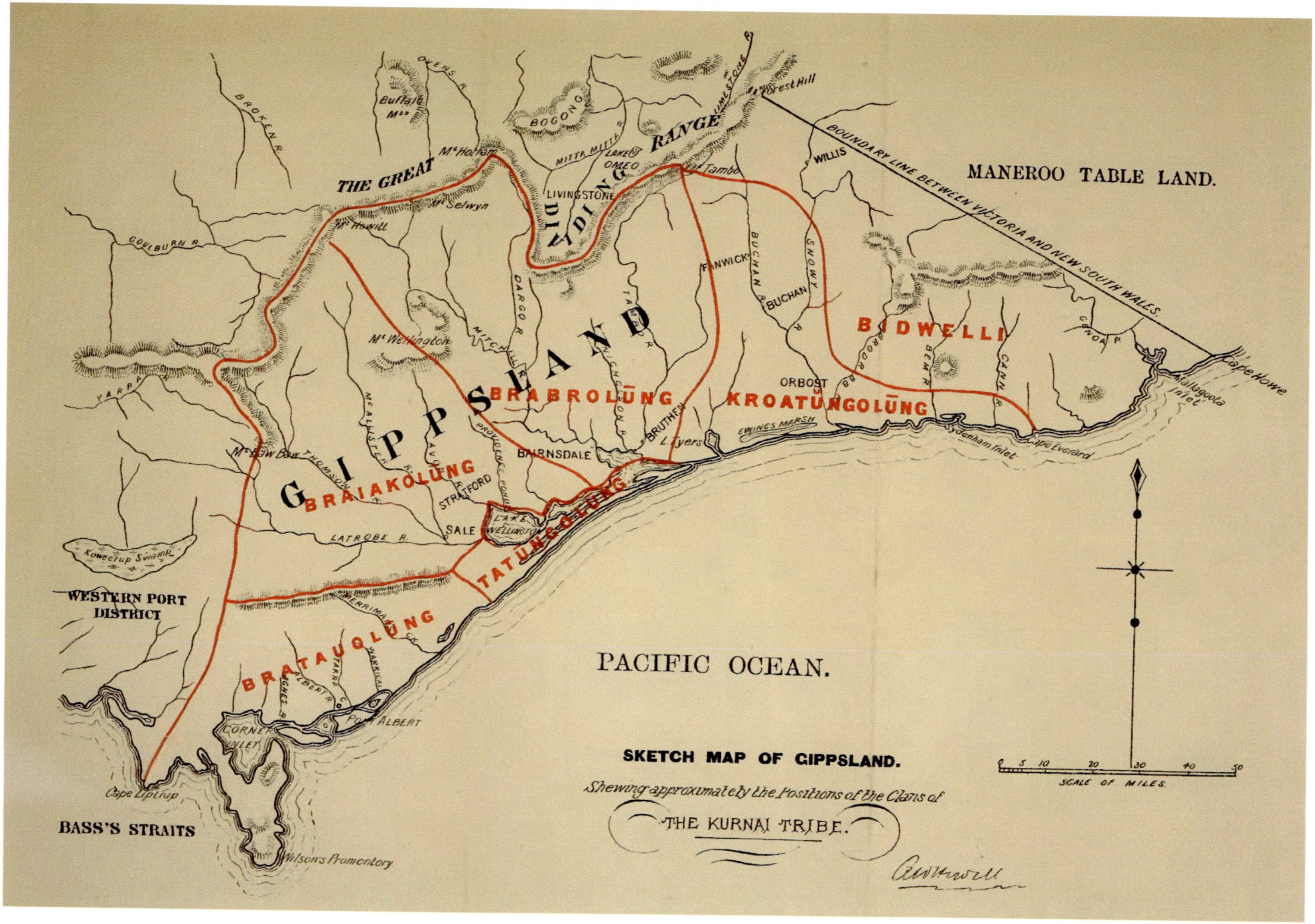

Figure 1.3. 'Sketch map of Gippsland, showing approximately the positions of the clans of the Kurnai tribe' (from Fison and Howitt 1880: 8).

Contact with outside groups was more fluid on the outer edges of the five clan territories. Cross-cutting clan affiliations, the GunaiKurnai were divided into two ancestral eponymous 'sex moieties' named after birds (totems), *yiirung* (Emu Wren, *Stipiturus* sp.) or 'elder brother' for the males and *djiitgun* (Superb Warbler, *Malurus cyaneus*) or 'elder sister' for females (Howitt 1904: 103, 148). Neither of these species could be injured, for to do so would be tantamount to injuring kin of that sex. 'Fights between the sexes on account of the killing of the brother or sister totem' were widespread (Howitt 1904: 148–149; Pepper 1980: 37). This highlights one among many ways in which GunaiKurnai Country and all within it were, and in many cases continue to be, socialised: fire damage to the landscape is also damage to GunaiKurnai society and culture, encompassing not just artefacts but also the animals and the plants. They are more than zoology and botany; they are part of the GunaiKurnai social world. Traditionally GunaiKurnai daughters took their mother's sex moiety and sons took their father's, with marriage taking place outside (i.e., exogamously to) one's local residential group (which consisted largely of close kin) and undertaken by arrangement, elopement or capture. *Yiirung* therefore married *djiitgun* or outsiders (*brajerak*), and residence was predominantly patrilocal (Fison and Howitt 1880: 199, 204, 227). GunaiKurnai men married GunaiKurnai women of the opposite sex moiety, including across clans and with *brajerak* especially from the east. GunaiKurnai men lived in their father's Country and women lived in their husband's Country, and men could, if marrying *brajerak* women, access their wives' Country to the east and to the north (Fison and Howitt 1880: 204). Accordingly, people also travelled across the landscape through such associations, connecting to different parts of the landscape through kinship affiliations and relationships, spreading material goods and cultural practices along the way, and creating and maintaining travel routes and various kinds of meaningful places in the process. Different parts of GunaiKurnai Country were culturally interconnected in various ways, and individual events in one area could affect GunaiKurnai or other Aboriginal groups further away, including where to burn the landscape to maintain 'healthy Country' according to ancestral and kinship affiliations and managerial rights. While some of the old ways are not now followed, others are, including relationships to clan lands and the managerial roles that these entail.

GKLaWAC HEALTHY COUNTRY, WHOLE-OF-COUNTRY PLAN

GKLaWAC has developed a Whole-of-Country Plan, towards the nurturing and maintenance of healthy Country that incorporates GunaiKurnai knowledge and educational opportunities for community members. Fire and its management is an important dimension of keeping Country healthy.

It is important to understand what the notion of 'Country' means in GunaiKurnai culture. Country includes the land, waters and sky and all the living and inanimate things such as the rocks and soils within it all as one. But it is also more than this: Country includes the ancestral presences, their past actions, the present peoples, and the relationships between them and with all the things that reside or pass through the GunaiKurnai landscape. GunaiKurnai management of Country includes the cultural ways of keeping the landscape healthy, as passed down from the Old Ancestors together with new ways that are embraced by GunaiKurnai Traditional Owners. GunaiKurnai Traditional Owners 'are guided by the spirits of our ancestors when we walk through this Country. ... We have a cultural responsibility to ensure that all of it is looked after. ... Our spiritual connection is something that cannot be seen, but nevertheless exists strongly in the places we walk and in the paths of our ancestors' (Gunaikurnai Land and Waters Aboriginal Corporation 2015).

The GKLaWAC Whole-of-Country Plan has direct relevance to the management of Country and wildfires:

> [It] has drawn heavily from the aspirations that our mob have expressed over many years. We have worked hard to be faithful to all of the work that was done before, and bring it into the new context in which we are now operating. We are now embarking on a fresh push to implement the things that our mob has cared about for a long time. (GunaiKurnai Land and Waters Aboriginal Corporation 2015).

The Plan is based on nine fundamental Principles:

> **We have cultural obligations.** It is our inherent responsibility to look after Country—to heal the damage of the past and protect it for future generations.
>
> **Everything is connected.** All of our Country is linked. There is no separation between our landscapes, waterways, coasts and oceans, and natural and cultural resources. All are linked and bound to our people, law and custom.
>
> **Every bit matters.** We understand the need to prioritise limited resources to where important values are under threat, but every part of our Country remains important to us. Our values exist even when you can't see them—whether they are under water, deep inside caves, covered with vegetation, they are still important to us.
>
> **Don't wait until it has gone.** When you lose a site, it's gone forever. We need to act now to prevent any further loss of environmental or cultural values.
>
> **Look at what was there before.** When we are healing and restoring degraded landscapes, we should try to put back the plants and animals that used to be there.
>
> **Sustainable use.** Our approach to managing Country is to balance resource use with conservation—they are all part of the same. Take only what you need—leave some for others.
>
> **Seek collective benefits.** We use our resources for the benefit of our mob rather than seek individual gain.
>
> **We have the right to be on our Country.** Traditional Owners should not be restricted in accessing our traditional Country. At the same time, we should have the right to restrict access to others who disrespect and damage our sensitive areas.
>
> **Our traditional knowledge is valuable.** Our traditional practices and approaches sustained the land for thousands of years. Our Country should be managed in harmony with our traditional ways. We need to take the time to understand what natural and cultural heritage exists out on Country. It can't be managed properly if we don't know what is there. (Gunaikurnai Land and Waters Aboriginal Corporation 2015: 16).

In this context, it is important to note that GunaiKurnai cultural heritage sites—archaeological sites and story places—are non-renewable ancestral places. They are the GunaiKurnai 'history books' written in the landscape. Through time, as more and more sites get damaged or destroyed through disasters such as wildfires and the expanding footprint of developments such as roadworks, urban growth and the like, progressively fewer of the old ancestral sites survive; GunaiKurnai cultural heritage sites are a diminishing 'resource'. It is thus especially incumbent on land managers to treat cultural places carefully, and in the event of widespread calamities such as wildfires, to retain a sharp awareness of the special nature of Aboriginal places during fire-management and fire-fighting planning to recovery.

After a wildfire, the GKLaWAC Bushfire Recovery Crew works across the fire's footprint area, monitoring the impacts of wildfire on cultural heritage sites and the broader landscape, towards the management of the fire's impacts such as the recovery of animal and plant species that are significant to the community. Such management practices are also a means of cultural learning on Country. As GKLaWAC points out:

> Country heals us and connects us to our ancestors, our culture and our history. But our mob cannot be healthy when Country is sick which is why it's been so important to get community out and involved in bushfire recovery—reading and healing, connecting and sharing knowledge. (https://gunaikurnai.org/our-country/bushfire-recovery/).

GKLaWAC AND THE VICTORIAN TRADITIONAL OWNER CULTURAL FIRE STRATEGY

Articulating with the GKLaWAC Whole-of-Country Plan towards healthy Country is the *Victorian Traditional Owner Cultural Fire Strategy*. Its purpose is to 'reinvigorate cultural fire through Traditional Owner led practices across all types of Country and land tenure; enabling Traditional Owners to heal Country and fulfil their rights and obligations to care for Country'. 'Cultural fire' refers to the knowledge and practices of intentionally burning parts of Country towards its short-term to long-term management. Cultural burning is done at particular times of the year, under appropriate environmental and social conditions, by the rightful Traditional Owners and sometimes with their approved affiliates for that land, and using fire technologies that are apt for the job at hand. 'Cultural burns are used for cultural purposes—they are not simply about asset protection. Cultural burns protect sites and clear access through Country for cultural uses—hunting, access to fish traps, ceremony etc.'. (Victorian Traditional Owner Cultural Fire Knowledge Group 2019).

With these aims in mind, the *Victorian Traditional Owner Cultural Fire Strategy* is based on six Principles:

Principle 1. Cultural burning is right fire, right time, right way and for the right (cultural) reasons according to Lore. There are different kinds of cultural fire practices guided by Lore applicable across Victoria's Countries.

Principle 2. Burning is a cultural responsibility. Traditional Owners lead the development and application of fire practice on Country; the responsibilities and authority of Traditional Owners are recognised and respected.

Principle 3. Cultural fire is living knowledge. Aboriginal fire knowledge is shared for continual learning and adaptive management. Traditional Owners will work together on each other's Country to heal Country and guide practice development. Knowledge and practice are shared.

Principle 4. Monitoring, evaluation and research (MER) support cultural objectives and enable adaptive learning. MER will be used to build a body of evidence that allows cultural burning to occur and grow.

Principle 5. Country is managed holistically. Traditional Owners manage Country holistically to address multiple values and objectives, healing both Country and culture. Partnership arrangements and management objectives are tailored to each regional and cultural landscape context. This includes analysis of the tenure, regulatory and operational arrangements to support cultural fire application, other beneficial Indigenous management practices, together with a process of learning to continuously improve planning, management and action.

Principle 6. Cultural fire is healing. There are substantial positive impacts to Traditional Owner wellbeing and confidence through providing access and authority to practice on Country.

GunaiKurnai cultural burning is about caring for Country, and helps restrict and manage a landscape that is less at risk of uncontrolled wildfires, for example by burning dry undergrowth and leaf litter that can act as kindling in Gippsland's hot summers, or burning in relatively small areas or 'patches' so that different parts of the landscape retain differential patches of growth. Cultural fire is a tool for managing Country (Victorian Traditional Owner Cultural Fire Knowledge Group 2019). Accordingly, the *Victorian Traditional Owner Cultural Fire Strategy*, embraced by GKLaWAC, has four major Objectives:

a) For Traditional Owners to develop and lead on-Country cultural burning pathways.
b) To build Traditional Owner governance and capacity in cultural fire knowledge and practice.
c) To improve landscape management through collaborative practice, to heal Country, and build community and landscape resilience.
d) To develop and strengthen institutional frameworks that support cultural burning practices.

While this volume is concerned mainly with wildfires, it is aligned both with the *Victorian Traditional Owner Cultural Fire Strategy*'s four Objectives, and with GKLaWAC's aspirations to build respectfully shared knowledge and research that can add positively to the GunaiKurnai management of GunaiKurnai Country (including to reduce current and future negative impacts of wildfires), as expressed in the Whole-of-Country Plan.

PART 1

Background to Fires and Cultural Burning on GunaiKurnai Country

Previous page: GunaiKurnai Ranger, Harley Finn managing Country using fire at Lake Tyers State Park, East Gippsland. Photograph by Jessica Shapiro, courtesy of GunaiKurnai Land and Waters Aboriginal Corporation.

Chapter 2

Wildfires: Characteristics, Drivers and Impacts on Cultural Sites

Grant Williamson and Jessie Buettel

Australia is a continent shaped over many millennia by fire. The flora and fauna, from tropical savannas, to the arid interior, to the wet forests of southeastern Australia, have lived and evolved under fire, and many have traits that let them tolerate and thrive under this disturbance. People have also lived alongside landscape fire, and have used it as a tool for many thousands of years (e.g., Adeleye *et al.* 2021; Haberle *et al.* 2010). However, not all fires are created equal; cool burns in the cooler months have very different behaviours and impacts on Country than the intense, difficult-to-control wildfires that dominate the summer months.

WILDFIRE CHARACTERISTICS AND BEHAVIOUR

Many ecosystems in Australia rely on fire, even intense wildfires, to function and remain healthy. Fires release nutrients into the soil (Chambers and Attiwill 1994), promoting new plant growth, and trigger germination in many species (Penman *et al.* 2008; Santana *et al.* 2018). Some systems, like wet forests, only burn very rarely under the most extreme conditions in high-intensity fires, but this disturbance is required to allow establishment of the next generation of trees (Waters *et al.* 2010). However, wildfires that are too intense, or too frequent, for a given ecosystem, can disrupt these processes of regeneration and result in a loss of species and a shift to a new ecosystem type (Bowman *et al.* 2014). Conversely, a long-term lack of fire in some systems can induce an ecological state shift (Russell-Smith *et al.* 2004; Baker *et al.* 2021). Plants that require fire for regeneration die out and are replaced by fire-sensitive species, with further implications for the associated fauna.

An important characteristic of a wildfire, compared to a prescribed or cool burn, is the **intensity** of the fire. Intensity refers to the energy released by the fire, the amount of heat it produces as it consumes the fuel, and this is often closely related to the height of the flames (Alexander and Cruze 2011). Wildfires are more intense than cool burns, producing greater heat (which can damage infrastructure and cause greater risk to people) and higher flames. Intensity describes the *energy* released by the fire, and is often measured and expressed in units of kW/m radiating from the fire-line. This fire intensity, which translates to flame length and temperature experienced in front of the fire, varies with fuel type and weather conditions (Table 2.1). There is also a relationship between the intensity of the fire and its **severity**. Unlike intensity, which refers to the heat and energy of the fire-line, severity is a measure that describes impact of the fire on the environment, in terms of the change of the above- or below-ground loss of organic matter (Keeley 2009). Severity illustrates another difference between wildfires and cool burns; wildfires tend to have higher severity, consuming more of the available fuel and producing flame heights that can damage or remove higher layers of

Table 2.1. Fire-line intensity and fire regime intervals for common vegetation types of temperate southeastern Australia (after Murphy *et al.* 2013).

		Intensity (kW/m)		
Vegetation	**Surface Fuel Type**	**Typical**	**Extreme**	**Typical Interval (years)**
Tall/Wet Eucalypt Forest	Sclerophyll Litter	5000–10,000	>50,000	20–100
Dry Eucalypt Forest	Sclerophyll Litter	1000–5000	10,000–50,000	5–20
Eucalypt Woodland	Tussock Grass	100–1000	1000–5000	5–20
Temperate Rainforest	Rainforest Litter	0–100	100–1000	>100
Heath	Sclerophyll Litter	1000–5000	10,000–50,000	20–100
Pasture/Cropland	Tussock Grass	100–1000	1000–5000	5–20
Mallee	Sclerophyll Litter	1000–5000	10,000–50,000	20–100

the forest canopy, while cool burns have a lower severity, impacting only layers of vegetation close to the ground, and often in a patchy pattern.

Rate of spread is the speed at which a fire front moves, and is driven by the fuel, the shape of the landscape and, importantly, weather conditions. Rate of spread is often greater in a wildfire and contributes to the difficulty of fire control and suppression (Sharples *et al.* 2009). In Australian forests, under extreme weather conditions when flames reach the canopy, wildfires can spread very rapidly. Fuel, and the continuity or connectedness of that fuel across the landscape, also plays a role, with wildfires having been observed to slow down upon reaching areas that have undergone recent prescribed burns.

As well as these characteristics that define how a fire burns, we can also characterise wildfires in terms of their **fire regime** (Murphy *et al.* 2013; Pausus and Keeley 2009). One element of this is fire frequency or interval; how often has an area burnt throughout history, and what is typical for that vegetation type? This varies greatly across Australia, with tall, wet forests having a fire return interval in the scale of centuries, while savanna grasslands in northern Australia can burn almost every year. Another feature of wildfires, compared to cool burns, is their typical size. Due to the high intensity and rate of spread under hot, dry weather conditions, and the difficulty in controlling them, wildfires can burn extremely large areas over multiple days. Cool burns, on the other hand, tend to cover small areas in a patchy mosaic, and due to the cool, moist conditions overnight, usually extinguish themselves in the evening.

A range of factors drive the occurrence, behaviour, intensity, and ecological severity of wildfires (Figure 2.1). **Fuel** refers to the organic matter that burns in a fire (Hollis *et al.* 2015); this can include surface fuels like bark, leaves and dried grass on the ground, through to live shrubs and bushes, heavier logs, and the forest canopy itself under extreme conditions. The types of fuel present in different vegetation drives fire behaviour, such that wildfires in open grassland have different intensities and fire regimes than those in dense, closed forests. Fuel is produced by the vegetation present, so while a fire consumes fuel, it will build back up over time after the fire (Olson 1963). The rate at which this occurs is different in different environments; grassy systems will regrow the grass rapidly in the next growing season, while

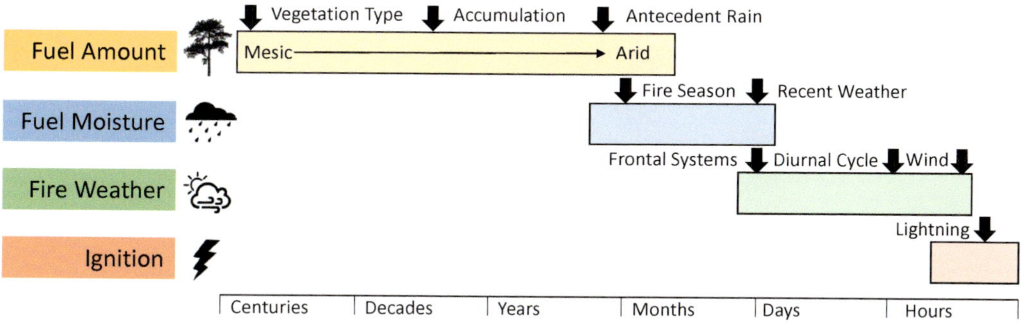

Figure 2.1. We can view fire occurrence as driven by four switches that all need to be turned 'on' for a fire to occur: fuel needs to be present; it needs to be dry; there needs to be appropriate weather conditions for fire to spread; and there needs to be a source of ignition, natural or human. These various 'switches' operate over different time scales; fuel may take decades to build up in a system, whereas ignition is instantaneous.

dropped bark, twigs, branches and leaves build up more slowly over many years in southeast Australian forests but can reach much higher loads. The primary tool we possess to alter the intensity of wildfires is through managing the fuel load (Fernandes and Botelho 2003), the mass of fuel that is available to burn in a fire. Cool or prescribed burns consume some of that fuel under moderate weather conditions, so there is less available to drive wildfires under hot, dry conditions.

In the short term, once a fire has started, weather conditions are a major driver of fire behaviour, and people have little control over them (although human-driven climate change, where it produces hotter and drier conditions, is expected to increase fire risk). A useful summary of the weather elements that drive fire can be found in the measurements used to calculate the Forest Fire Danger ratings, as commonly reported in daily weather forecasts over the summer. Wind speed, temperature, relative humidity, and a drought factor (McArthur 1967) that describes the amount of moisture in fuels are combined to determine the fire danger, which in turn gives an indication of expected rate of spread, intensity, and flame height (Figure 2.2).

The amount of moisture in the fuel is related to recent weather conditions, with precipitation introducing moisture into the fuel, and evaporation on warm, dry days drying it out. However, different types or sizes of fuels respond to external moisture on different time scales. Very fine fuel components, like leaves and twigs, will become saturated easily when it rains, but then dry out rapidly when the weather is warm; but larger bulky fuels, like fallen logs, may take months to dry out after rain to become flammable, and may remain dry enough to burn after moderate amounts of rainfall occur (Matthews 2013). Dry fuels, high fuel loads, high temperatures and low humidity all increase the amount of fuel available to burn, the rate at which it burns and the energy released, and high wind speed can increase the rate of spread and ensure the fire is fed with oxygen. In southern Australia, extreme fire danger days are often characterised by strong, hot, north or northwesterly winds carrying dry air from the interior of the continent. Often these conditions are followed by a cool change that brings a sudden shift in wind direction, spreading fires out suddenly in new directions. Under these conditions, wildfires can be difficult to impossible to control.

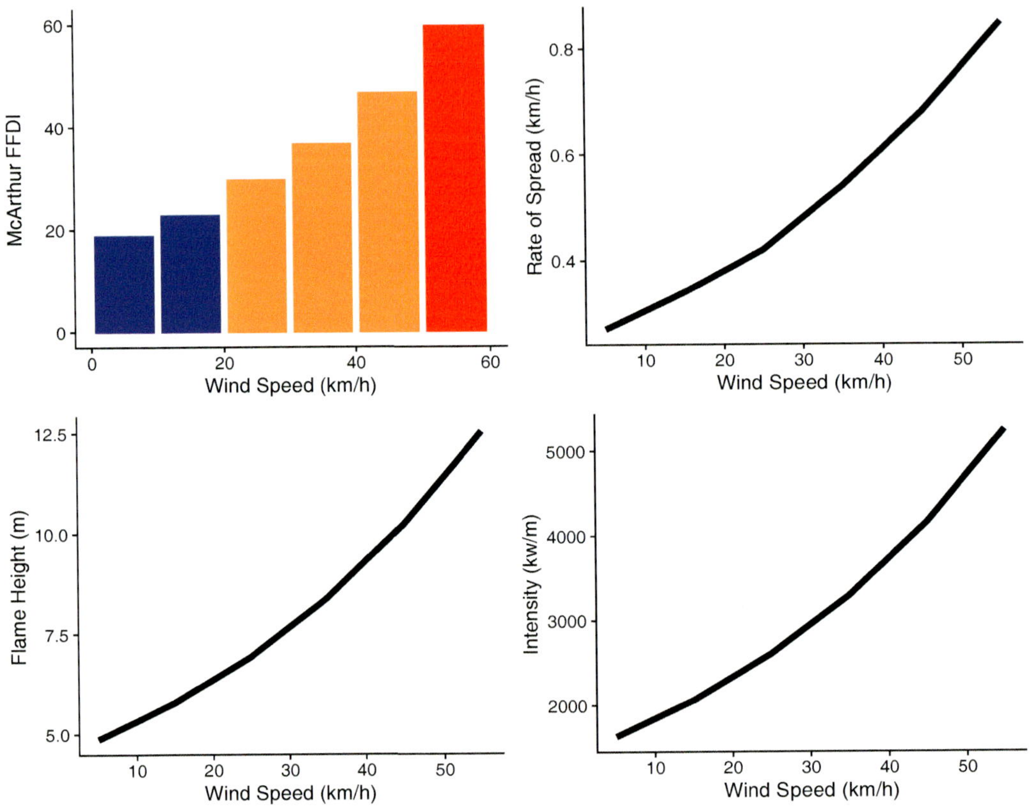

Figure 2.2. Influence of a weather variable (wind speed) on fire behaviour. Graphs are plotted for a forest under moderate drought (Drought Factor of 8), 30 °C temperature, 15% relative humidity, and a fuel load of 12 tonnes/hectare, typical for southeastern Australian dry forests in summer. The colours on the bar graph represent fire danger index levels: Blue = High; Orange = Very High; Red = Severe.

The landscape and terrain also drive wildfire behaviour. Fire tends to travel uphill faster and with a greater intensity than when travelling downhill, as flames are better able to reach unburnt fuel ahead of the fire when it is upslope of the front. North-facing slopes tend to be drier and warmer due to their greater exposure to the sun, which drives both different vegetation and more intense fire behaviour, than on moister south-facing slopes. Sheltered valleys with watercourses and wet, riverine vegetation often make natural barriers to wildfire spread, although under severe drought conditions, as were experienced in the 2019–2020 Gippsland fires (sometimes referred to as the 'Black Summer' fires), even these areas may be flammable and stop acting as barriers.

For fire-fighting personnel on the ground, the height of the flames provides a good indicator of the heat they will be facing, as it is strongly related to the intensity or energy of the fire. Research has shown that firefighters in protective clothing may be able to tolerate a radiant heat of $7kWm^{-2}$, which for high-intensity wildfire in extreme weather conditions, with flames approaching 25m in height, a safe distance of 100m from the flames can be maintained at a minimum (Figure 2.3). For cool burns under mild weather conditions, where flame heights are low and less fuel is available to burn, the safe distance is much shorter.

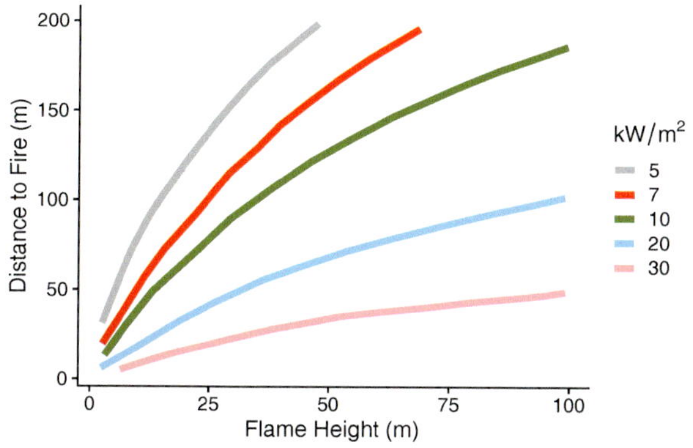

Figure 2.3. Relationship between flame height, distance from the fire-line, and heat intensity (after Butler and Cohen 1998). The red line shows the limit to prevent injury while wearing fire-fighting apparel. The graph indicates that, for example, if someone is 50m from a fire, and the flames are 50m high, the person is below the red line and therefore at risk of injury. If they are 150m away from the fire and the flames are only 25m high, then they are above the red line and relatively safe.

IMPACTS OF WILDFIRES

Wildfires impact human settlements and infrastructure, given their high intensities and difficulty of control, and land-use changes that are seeing more people living on the urban fringe and on properties enveloped by fire-prone forests, are shifting the profile of fire risk across the country (Hughes and Mercer 2009; Price and Bradstock 2013). Other land-use changes, such as the conversion of rarely-flammable wet forests to younger, more open plantations of different species, can also change the fire regime and the risk of fire.

One of the important secondary effects of severe wildfires, beyond immediate impacts on the flora and fauna, are changes to the land surface through erosion (the transport of soil and rocks) (Shakesby *et al.* 2007) and denudation (the overall lowering of the land surface). The consumption, by the fire, of surface leaf litter leaves the soil surface exposed, while the death of live shrubs results in a loss of roots holding the soil together. This effect is particularly pronounced when fires are followed, in subsequent months, by significant rainfall events (Alexandra and Finlayson 2020), before the vegetation can recover. The immediate erosion risk particularly impacts drinking water catchments, with the potential for soil and ash from the fire to wash into rivers, lakes and reservoirs.

Severe wildfires have three main impacts on the preservation and maintenance of cultural artefacts and layers in an archaeological site, the first through the direct impact of heat from the fire on the material, the second through the erosion and denudation processes detailed above, and the third as the result of human responses to the fire.

Soil is, generally, an excellent insulator for heat. While above-ground flame temperatures in the burning fuel and vegetation may range from over 1000 °C at the base of the flame to 200–300 °C at the tip of the flame, only a few centimetres under the soil temperatures may be quite survivable for many organisms (Bradstock and Auld 1995). Through these insulating properties, a soil seed bank that enables vegetation reestablishment is able to survive. However, despite temperatures being significantly lower underground than above ground, the temperatures are still elevated and can have an impact on artefacts and materials. The degree of impact depends on the depth at which the materials lie, with heat at the surface being much greater than at depth (Table 2.2). Penetration of heat into the soil is driven to some extent by the intensity of the fire itself, based on weather conditions and fuel, although this effect appears less important than the effect of the residence time (Wotton *et al.* 2011) of the fire—how long it continues to burn at a given location before fuel is consumed and the fire is extinguished. Fires in light, grassy fuels tend to move rapidly, burn out the fuel at a given location quickly, and have a lower temperature than fires burning through coarse woody debris, where heavy logs may remain burning for hours after the main fire-front has passed, significantly heating the soil underneath. Moist soil, which is more likely to be present during cool-season burning, may also act as a better thermal insulator than extremely dry soil.

Many studies of below-ground temperatures from fire are based on data collected in the United States and Canada. However, these data are of little relevance to Australia. Fire-prone North American ecosystems tend to be dominated by conifers (pines) which leave a thick layer of *duff* consisting of decaying pine needles and related organic matter (Figure 2.4). This layer can continue to burn hot for a long time after the main fire-front has passed, and is close to the soil surface, so its attributes have to be considered when determining underground heating. Australian ecosystems often lack this dense organic duff layer, with the surface fuel being dominated instead by more aerated *Eucalyptus* leaves and bark, which burn and extinguish more rapidly. It would be beneficial for more data on soil temperatures under varying vegetation types, fuel loads and fire conditions to be collected in southeastern Australia, including in GunaiKurnai country. This can be accomplished by installing thermocouples at defined depths under the soil surface (0cm, 2.5cm, 5cm and 10cm) prior to prescribed or cool burning and recording temperatures as the fire front passes. A limitation to this methodology, however, is the difficulty in measuring more extreme wildfires, as their timing is unknown before they happen, rushing to install equipment ahead of the fire-front is dangerous, and there is difficulty in maintaining the recording equipment in the extreme temperatures of the fire-front. An alternative approach is to install aluminium tags below the surface marked with temperature-sensitive paints (Rebbeck *et al.* 2006). A variety of paints that change appearance at different set temperatures are available, which can help determine maximum temperatures reached as the fire passes.

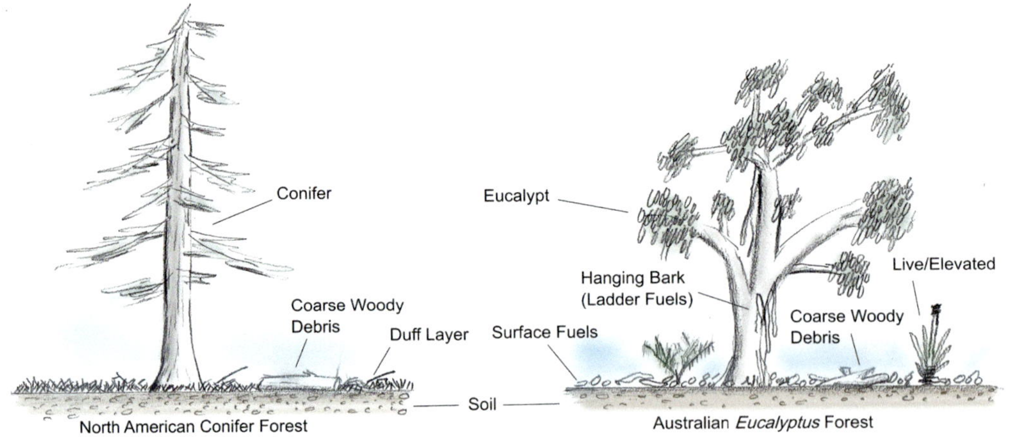

Figure 2.4. Comparison of North American fuel structure (left) and Australian eucalypt forest structure. North American forests, being dominated by conifers, have a dense duff layer, while eucalypt forests have a surface layer dominated by bark and leaves. Hanging strands of bark in some *Eucalyptus* species can help transfer fire to the canopy.

Table 2.2. Field and experimental measurements of soil temperature at various depths. Temperature declines sharply with depth over just a few centimetres, and even severe fire differs little from prescribed fire at 5cm depth.

	Depth					
	In litter	**0cm**	**1cm**	**2cm**	**3cm**	**5cm**
1) **Prescribed fire (*E. pauciflora* forest), ACT** (Raison *et al.* 1986)	600 °C	450 °C		54 °C		42 °C
2) **Moderate fire, Blue Mountains NSW** (Bradstock and Auld 1995)		150 °C		50 °C		36 °C
3) **Experimental 25 kW/m^2** (Enninful and Torvi 2008)			141.8 °C		50.6°C	35.5 °C
4) **Experimental 75 kW/m^2** (Enninful and Torvi 2008)			449.6 °C	102.2 °C	53.0 °C	

The actual direct impacts of the heat on surface or buried artefacts and archaeological layers depends on the physical material and properties of the artefacts and layers. In the case of wooden artefacts or those made from other flammable biological materials, complete consumption or scorching by the fire is possible, particularly where the items are close to the surface, while for artefacts of stone, shell or bone, the passage of fire can result in cracking, sooting or oxidation of the material.

Erosion and denudation after the fire has passed, often associated with significant rainfall events, may uncover and expose previously buried artefacts and layers of sites, leaving them vulnerable to increased degradation by the elements or future fires, or possibly to looting. Disturbance of the soil from fire-killed trees falling may shift or bury materials. Artefacts may also be transported through erosional processes away from their original locations, most likely down-slope or towards water courses. In such cases, stratified ancestral sites with great archaeological potential could be entirely destroyed through post-fire erosional events. This is particularly problematic because the ability of a site to tell the story of the ancestors largely depends on an ability to not be disturbed, to remain in situ, so that each layer tells its own story. Conversely, the erosional processes further up-slope may bury or increase the soil depth above material in down-slope or valley areas. Overall, the movement, exposure or burial of cultural materials through these processes may complicate stratigraphic understanding of a sediment sequence and history of artefacts following a fire. These are all, to a large extent, processes that are expected to occur in an exposed environment that wildfire, particularly severe wildfire, will accelerate.

The final factor impacting the state of artefacts and sites during and after a wildfire, and one often more important than the fire itself, is the human one (e.g., Swete Kelly and Phear 2004). Actively fought wildfires are zones of high human activity, both on foot and by vehicle, where the immediate priority is protecting lives and assets, biased towards buildings and infrastructure, rather than surface and buried GunaiKurnai cultural materials that may be less easy to see. During fire-fighting operations, areas can become trampled, and earthworks may shift significant quantities of soil, shifting the location and depth of materials on the surface or underneath the soil. This disturbance may continue after the fire-front has passed during 'blacking out' operations where fire agencies identify and extinguish smouldering logs and debris. Finally, as highlighted above, the loss of vegetation cover and exposure of shallowly buried material may enable interference and theft by people visiting the burnt area.

Simply as a function of the lower intensity of prescribed fire, with less heat penetrating the soil, lower consumption of fuel (and particularly of coarse fuels) and less firefighting activity on the ground, prescribed or cool burns would be expected to have less impact on the preservation of artefacts, archaeological sites and their stratigraphies than more extreme summer fires. Changes in climate, particularly warmer temperatures in the traditional cool burning season (spring and autumn) have the potential to shorten the window available for prescribed burning, and a projected increase in extreme weather events, including heat waves and intense rainfall events, may pose increased risk for the preservation and retention of cultural materials impacted by fires.

CONCLUSION

In summary, while wildfires are undeniably important for ecosystem health, if they are too intense or too frequent, they have the potential to disrupt and stymie the regeneration of vegetation, change the land surface (through erosion and denudation) and ultimately cause the loss or damage of non-renewable cultural sites (see especially Chapter 12). Exploring and understanding how the spatial pattern of fire behaviour has changed through time within GunaiKurnai Country can give us insights into the pattern of risk for cultural sites. When paired with an understanding of the underlying drivers and characteristics that lead to more frequent, large-extent, hot and intense events and thus more widespread negative impacts, such information provides critical details towards the protection and management of cultural sites, now and into the future (see Chapters 12–15).

Chapter 3

Accounts and Memories of Landscape Burning Practices in Gippsland

Seumas Spark

The historical and anthropological records tell of Aboriginal burning practices in Gippsland, including those of the GunaiKurnai, but this information rarely contains explicit details, and little of it comes from members of Aboriginal communities themselves. (The name 'GunaiKurnai' and its variant spellings is absent from most of the records on which this chapter is based; it is used here in its present form for uniformity and as advised by the GunaiKurnai Land and Waters Aboriginal Corporation). As the environmental historian Stephen J. Pyne notes of Gippsland in his history of fire in Australia, 'there are few solid references to routine Aboriginal burning' (Pyne 1991: 212). The voices of those who have cared for and burned the land for thousands of years are mostly absent from the written historical record.

Nonetheless, there is value in this written record of Aboriginal burning practices in Gippsland, even if those observations are sometimes now known to be misguided or wrong. A close reading can provide insights. In their *Report on the Physical Character and Resources of Gippsland*, published in 1874, Skene and Brough Smyth wrote: 'Nature in these regions has not been interfered with by man; and yet there was something almost artificial in the aspect of the hollow around the Diamantina Springs. The richly foliaged shrubs seemed to have been set in their places in obedience to rules of art ...'. Skene and Brough Smyth's description hints at their having missed something, though it is unlikely that either considered the possibility of a mosaic landscape created through the careful application of fire by the area's Aboriginal inhabitants (Skene and Brough Smyth 1874: 39).

This chapter reviews the literature and knowledge on Aboriginal burning practices in Gippsland. It focuses on 19th and 20th century literature, broadly chronologically and beginning *c.* 1840 with the non-Aboriginal (colonial) settlement of Gippsland. The literary record shows that White settlers in Gippsland learned that fire played a pivotal role in taming and managing the land; a recognition that connected them, in one small way, to the Aboriginal peoples of Gippsland. In the absence of mechanised machinery, fire was the most effective tool for clearing land, and necessity dictated that newcomers to Gippsland wield it. But there the understanding usually ended, for most newcomers did not learn Aboriginal ways with fire: when, how and what to burn. In the hands of the first White settlers, fire was a fierce and unpredictable weapon rather than a delicate instrument applied to Country with care.

Here I discuss the central importance of the work of Alfred Howitt, scientist, explorer and author, and show how much his pioneering scholarship informs the written historical record. I discuss also the importance of recording oral histories, including with non-Aboriginal people, to add to what is known of Aboriginal burning practices. There is evidence that non-Aboriginal cattlemen and women in Gippsland and southern New South Wales have long followed the advice of Aboriginal friends and employees, burning when told to do so. Sometimes—usually when it was a matter of livelihood—White men and women knew to defer to Aboriginal knowledge. These (hi)stories are reflected in oral reminiscences, and sometimes in maps that show fire-managed paths and clearings, but rarely in the written record. Capturing these (hi)stories would add much to knowledge of Aboriginal burning practices in Gippsland, including in recent decades. Aboriginal burning in Gippsland has been practised repeatedly since White settlement in the 1840s, and sometimes for the purposes of non-Aboriginal people.

In preparing this chapter, a detailed search of newspapers was conducted via the Trove database. This unearthed useful results about Aboriginal burning practices, though fewer than expected. Mentions of GunaiKurnai burning in newspapers tend to be confined to book reviews—articles on Howitt's scholarship, for example.

Two key principles inform this chapter. The first is that the best way to increase knowledge of Aboriginal burning practices in Gippsland is to talk with Aboriginal people. That information could then enter the literary record, given necessary permissions. All that follows here is written with this principle in mind.

The second principle is that this chapter does not engage with the argument advanced by some commentators that the extent of Aboriginal burning practices in Gippsland before White settlement has been overstated or, even, fabricated. Often such ideas are motivated by politics rather than evidence. This chapter instead proceeds on the understanding that the GunaiKurnai and other Aboriginal groups fired the land regularly. Aboriginal and non-Aboriginal peoples of Gippsland know this, so too the vast majority of scholars—historians, archaeologists, anthropologists, ecologists, botanists and others. Indeed, this understanding is common to literature in the humanities, social sciences, and the pure sciences. For example, Gell *et al.* (1993) used scientific methods to show that Aboriginal people were burning country at Tea Tree Point in East Gippsland long before colonial contact.

This section begins by presenting information collated from a series of interviews conducted with GunaiKurnai individuals in early 2021. In these interviews, undertaken as part of a separate GunaiKurnai Land and Waters Aboriginal Corporation (GKLaWAC) project, interviewees were asked questions about their knowledge and use of fire. The respondents offered information on the basis that GKLaWAC would ask their permission before sharing it further.

THE GUNAIKURNAI AND FIRE

In January–February 2021, I undertook a number of interviews with both GunaiKurnai and non-GunaiKurnai Aboriginal residents in GunaiKurnai Country about Aboriginal landscape burning practices in Gippsland. Aunty Phyllis Andy was interviewed on 20 January and 24 February 2021; Cheryl Drayton on 21 January 2021; Terylene Hood on 25 February 2021; Uncle Russell Mullett on 11 and 25 February 2021; Cathy Thomas on 12 February 2021; and Aunty Glenys Watts on 12 February 2021. Aunty Phyllis has lived on GunaiKurnai Country all her life but is not a GunaiKurnai woman. Her mother was from Wotjobaluk Country near Dimboola in western Victoria. Aunty Phyllis is grateful to the GunaiKurnai for allowing her to live on their Country and share in its beauty.

When I asked GunaiKurnai individuals about the use of fire, most mentioned its vital role in keeping Country clean, healthy and safe. 'Clean' was a word used by several respondents. Cathy Thomas noted that small, knee-high, mosaic burns, conducted mostly in the colder months when ground is damp, ensured both the health of Country and its continuation as a source of bush-tucker and supplies. Cool burns benefit animals, and stimulate plants used by the GunaiKurnai for food and activities such as basket weaving. In the words of Aunty Glenys Watts, burning is an environmental strategy designed to protect an ecosystem. Aunty Phyllis Andy echoed this when she said that one reason for burning is to safeguard the future.

Both Aunty Phyllis and Terylene Hood remember that fire around Lake Tyers was once frequent, and not that long ago. Terylene spoke of her grandfather burning bush on a daily basis in the Lake Tyers area, while Aunty Phyllis recalled her father and other men burning Country at the Bluff in the 1950s and 1960s. Because this was men's work she watched from a distance, but Aunty Phyllis believes the men shared the cultural knowledge of what to do. She laments that such burning is far less widespread than it was. GunaiKurnai Country today is as dry as she can remember, a fact she attributes to the absence of the firestick: cool burns promote the growth of particular plants, keeping moisture in the ground. Speaking of the catastrophic fires in East Gippsland in the summer of 2019–2020, Aunty Phyllis said that she had 'never been so scared'. This wasn't Aboriginal fire, she told me, but the creation of those who ignore Aboriginal knowledge and choose not to burn. Western rules, noted Terylene, have greatly restricted GunaiKurnai management of Country.

The responses of those who were interviewed for this chapter certainly highlighted differences between Aboriginal and non-Aboriginal people in their understanding of fire and landscape. Uncle Russell Mullett and Cheryl Drayton mentioned that Country tells you when to burn, and that these signals from the land, rather than dates or arbitrary rules, should dictate when fires are set. Cheryl noted that in GunaiKurnai Country there can be six seasons a year, a concept that undermines the widespread practice of setting burns according to the less nuanced non-Aboriginal calendar of four seasons. Russell Mullett and Cathy Thomas mentioned also the importance of recognising and learning from the signals given by Country *after* a burn, cool or hot. What burned? What didn't? Which plants re-emerged, and when? Country both delineates Aboriginal approaches to setting fires, and shapes responses in their aftermath.

In her interview, Cathy Thomas emphasised that every fire is different and, as such, there is always a lesson to learn. After thousands of years of burning the land, GunaiKurnai continue to add to their knowledge of fire. Education is a protection against complacency, and a spur to action. As Russell Mullett observed, nature puts out fires, not people. Many non-Aboriginal people are yet to realise this.

Discovering a Fire-Managed Landscape

When the first non-Aboriginal, colonial settlers came to Gippsland, they saw a fire-managed landscape, though they didn't always recognise it as such. Angus McMillan, who was among the first White men to enter Gippsland, described 'beautiful open forest' and 'park-like land' on his exploratory travels in 1840 (cited in Watson 1984: 112–113; see also Bride 1898: 254–259; Fell 1978: 13–15). Paul Edmund de Strzelecki was more lyrical:

> The region eastward of the chain in the direction to Corner Inlet presents a totally different aspect. At the latitude 37°, or about the sources of the river Thomson, the spurs are less ramified, and of considerable height and length, shaping the intermediate ground into beautiful slopes and valleys, which ultimately resolve into a fine open plain, richly watered, clothed with luxurious grasses and fine timber, and offering charming sites for farms and country residences. Viewed from Mount Gisborne, Gipps Land resembles a semi-lunar amphitheatre walled from N.E. to S.W. by lofty and picturesque mountain scenery, and open towards the S.E., where it faces, with its sloping area the uninterrupted horizon of the sea. (de Strzelecki 1845: 63; see also de Strzelecki 1841).

Such views as described by de Strzelecki caught the eye of others, too, who saw Gippsland as a potential paradise for graziers. The historian and author William Westgarth described 'tracts of a beautiful open grassy country, resembling in some respects the rich and lightly timbered pastures of the western district' (Westgarth 1848: 28). In 1847, the surveyor Francis Peter MacCabe produced a 'Gippsland Rivers' map, with annotations indicating areas of 'luxuriant' grass, good 'pasturage' and 'open forest' (Figure 3.1). The map clearly depicts a fire-managed landscape, though MacCabe did not connect these park-like landscapes to Gippsland's Aboriginal inhabitants.

Nor was the link always apparent to those who had some appreciation of Aboriginal culture, such as the writer 'Tanjil' (Dow 2004: 133). Writing in 1886 of the 'settling' of Gippsland in the early 1840s, he described its grassy plains and open country, then added: 'Gipps Land in 1842 possessed all the features of a country in a state of nature. The white man had hardly made his mark, the black man's mark as yet predominated, though it did not amount to much, the remains of camps, trees notched in the act of climbing in the native fashion, near the lakes' (Tanjil 1886: 17).

Allan McLean (1840–1911) and his family were among the first White settlers in Gippsland, moving there in the early 1840s. McLean, who would later serve as Premier of Victoria and as a federal government minister in the early days of the Australian Commonwealth, grew up at Glenaladale. In a 1905 *Gippsland Times* article entitled 'Reminiscences of Early Gippsland', McLean remembered fire as a constant presence. In spring his family and other settlers lit fires to clear vegetation, and in summer they feared the destructive power of wildfires. He recalled Aboriginal uses of fire, but believed it was employed as a weapon rather than as a means for tending country: 'the blacks often started fires in order to burn us out' (Anonymous 1905: 3; see also RHSV, MS 000384, Box 125/7, Reminiscences of Allan McLean). There is evidence that Aboriginal people used fire as a defence against White colonists (see Cahir and McMaster 2018: 124–127).

Other White men and women did see the hand of people in the landscapes of Gippsland, and came to recognise that fire was the tool that Aboriginal people used to craft Country. The pastoralist William Brodribb travelled through Gippsland in 1841: 'The natives had burnt all the grass at Gippsland late in the summer. Heavy rains must have fallen before we reached there, in the month of March (Autumn). The whole country was very green. It had here the appearance of young cornfields; the young grass was about six inches high, and in places very thick'. It is tempting for modern readers to think that Brodribb was wrong about burning in summer, but this would be a contemporary conceit. GunaiKurnai Elder Russell Mullett

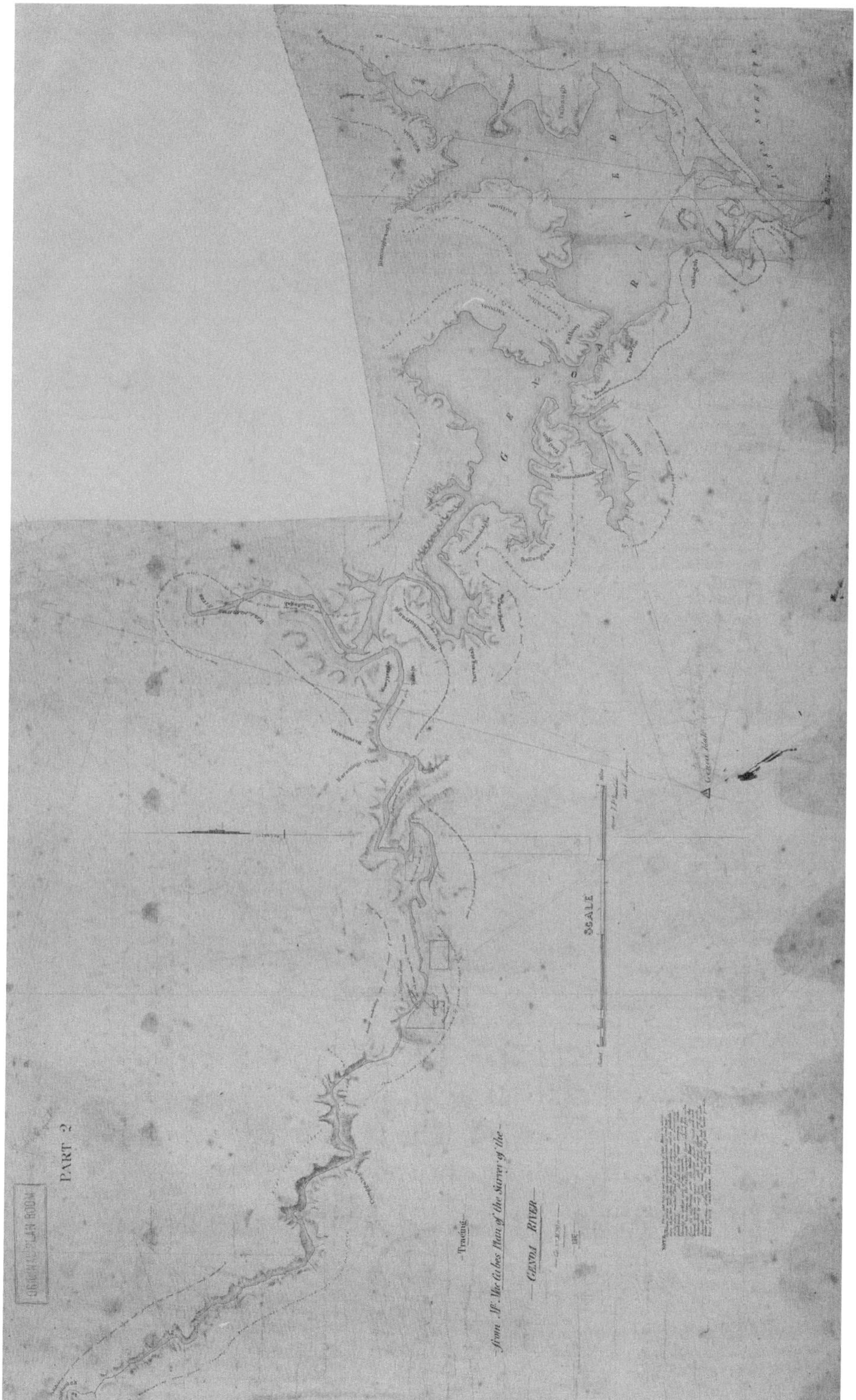

Figure 3.1. Francis MacCabe's 1847 map shows a fire-managed landscape. The map itself, and MacCabe's detailed annotations, make this clear (from Genoa River, GIPPS 54, Historical maps and plans, State Library of Victoria, Melbourne).

has confirmed that Aboriginal burning took place year-round, with the firing of landscapes depending, as always, on Country and conditions. In 1841 White settlement of Gippsland had not yet wrought the changes to country that would, in time, create the conditions for the firestorms now characteristic of Australian summers (Brodribb 1976: 24).

When travelling into Gippsland in 1844, George Augustus Robinson, Chief Protector of Aborigines for the Port Phillip district, connected the proliferation of messy scrub with the recent absence of Aboriginal burning (Clark 1994: 14). Later, near the Albert River in South Gippsland, he saw country 'that had been burned, which Robinson took to be evidence that natives had recently been in this country' (Clark 1994: 16). His travels in Gippsland and into the Monaro region of southern New South Wales helped him to connect Aboriginal burning practices with certain landscapes. Of the Monaro, Robinson wrote: 'It was a fine clear day in July when I first saw the Maneroo [Monaro] Country. The immense Downs with their undulating grassy surface stretched out before me as far as the eye could scan, a Park of great magnitude and beauty studded with copses of Banksia, Casuarinæ, Mimosa, Shrubs, and small belts of Eucalyptus' (Robinson *et al.* 1941: 14; see also Joubert 1876: 20). Such country, Robinson recognised, was the product of Aboriginal knowledge and practice.

The Reverend John Bulmer ministered at Lake Tyers from 1862 until the early 20th century. More so than most of his non-Aboriginal contemporaries, he was interested in GunaiKurnai culture and practice. He wrote:

> In hunting the kangaroo all the available men of the tribes went together. Each was armed with two or three spears, barbed with pieces of flint or in more modern times with broken glass and a *marriwan* for throwing them. They generally went in a very large circle, and gradually closed in, leaving a narrow opening for the kangaroos which were speared in passing. But in summer they set fire to a large tract of country and speared the animals as they were escaping from the fire. They also got many after the fire almost roasted enough for eating. (cited in Vanderwal 1994: 61).

Like Brodribb, Bulmer was a witness to the GunaiKurnai burning of Country in summer.

USING FIRE

The journalist Donald Macdonald's 1887 book *Gum Boughs and Wattle Bloom* includes a chapter on Gippsland, in which reference is made to Aboriginal uses of fire: '... for a fire-stick—the great native scavenger and disinfectant—was thrust amongst the thatch when the [Aboriginal] huts were abandoned. Indeed, there are traces of fire all along the slope that suggest a line from "Bush Ballad"— "All fire-flushed when forest trees redden on slopes of the range"' (Macdonald 1887: 78). Though few observers were as explicit as Robinson, Bulmer and Macdonald, 19th century literature shows that many of Gippsland's White settlers and visitors shared an understanding that Aboriginal people had used fire widely and regularly to shape the landscape. This understanding was often only vaguely formed, but it was there, as the literary record proves. White settlers knew also that fire might be made to work for them, specifically in clearing the landscape of scrub and bush. Ironically, some of the scrub they saw was due to Aboriginal people having been prevented from caring for Country and deprived of the chance to burn.

What the settlers often did not know was how to control fire. Caleb Burchett, born in 1843, settled at Poowong in South Gippsland in the 1870s. In a manuscript written in the early 20th century, he described clearing the land so that he could sow grass seed: 'This growth we called scrub and the first work undertaken was to cut this down and let it lie till after Xmas time, and then burn. If all things were favourable and a good burn was secured then the picking up of the timber left was comparatively light work and the fortunate selector would be said to have a "Good Burn". It was my misfortune to have almost every time a "Bad Burn"' (Burchett, no date).

In 1956, E.E. Straw expanded on Burchett's burning method:

> By the middle of December this portion of the block was strewn with piles of hazel scrub and dry undergrowth which would be ready to burn off by the end of summer, or earlier if a spell of hot weather produced favourable conditions. In this case, however, a heat wave and a long dry period made it possible to start the fire by the end of January. The great mass of fallen timber and debris had now become so dry and inflammable it only required a spark to touch off a great blaze. This was an anxious time for the settler. A good burn cleared up the block and minimized the work afterwards. A bad burn might postpone production for twelve months. Caleb therefore waited for the right weather—a hot day with a good wind was desired to fan the fire into an inferno. He postponed the event till the middle of February. This particular morning gave promise of a good start. Word was sent to neighbours, who assembled about 1 p.m. to help their fellow settler. They dispersed to different points of the clearing. At a given signal each one started the fire. They then made a torch of bark in order to light along the boundary. Once the fire had encircled the burn nothing more could be done. Everyone settled down to watch the seething mass of smoke and flame. Its concentration and intensity consumed everything within this 30 acres in a few hours. (Straw 1956: 6)

The concept of 'good' and 'bad' burns, and the idea that infernos were necessary, prevailed elsewhere in Gippsland. In 1920, a man named A. W. Elms contributed a chapter entitled 'A Fiery Summer' to a book on the non-Aboriginal settlement of South Gippsland. Scrub, he wrote, was burned 'In the Summer, on the hottest day available, and, if possible, with a strong wind blowing, the fallen scrub is set on fire, and if the burn is a good one, the fire burns up all leaves and small timber, leaving only the large saplings to be picked up and burnt off' (Elms 1920: 307). He added: 'Once started, there was no control over the fire, which might burn for weeks in trees and hollow logs, ready to spread afresh with wind or hot weather' (Elms 1920: 307; see also Wakefield 1970: 153).

The squatter Patrick Coady Buckley 'Set fire to tea tree scrub' in late November 1868. The month had been 'very dry', and two days after setting his burn, Buckley noted in his journal that there were 'Bush fires all around' (Buckley, 30 November 1868). These signs didn't stop him burning. A month later, on 29 December, he again 'set fire to scrub', though conditions had only worsened since November (Buckley, 29 December 1868). No rain had fallen.

I have included these examples in my review of the historical literature to show that non-Aboriginal residents of Gippsland also fired the land, prompted probably by direct and passive knowledge of Aboriginal burning practices. In his book *Dark Emu*, Bruce Pascoe mentions Jinoor Jack, an Aboriginal man, who explained to a White man how to burn in East Gippsland. Jinoor Jack mentioned a five-year burning cycle (Pascoe 2021: 167). The White man was Robert Alexander. Pascoe's source for this statement is a history self-published by Alexander's descendants: Jinoor Jack's knowledge, imparted in the 1850s or 1860s, was passed down through the Alexander family (excerpts available at https://www.towambavalleyhistory.webhive.com.au/alexanderfamily.htm). How many of Gippsland's White settlers received burning advice is unclear, though some certainly did: the concept of the five-year burning cycle was known to White settlers at different times and places elsewhere in Gippsland (Tonkin and Landon 1999: 208).

But there were crucial differences. The aims of the White settlers who burned the land were often different from those of the GunaiKurnai, their methods far less skilled and safe, and the results sometimes catastrophic (see Watson 2016: 11). The disruption caused to Country by the arrival of White settlers in Gippsland either made possible, or at least greatly exacerbated, the 1851 and 1898 wildfires. The science that informed cool burns, as practised by the GunaiKurnai and other Aboriginal peoples in eastern Victoria and southern New South Wales, was not generally understood by White settlers (Watson 2016: 12–13, 75). But they had learned that fire could be an ally. In this regard if nothing else, Aboriginal burning knowledge, corrupted and misunderstood though it often was, can be glimpsed in such writings as those of Burchett, Elms and Buckley.

ALFRED HOWITT AND HIS SCHOLARSHIP

Much of what has been written about Aboriginal burning practices in Gippsland derives from the work of Alfred Howitt (1830–1908), explorer, scientist and anthropologist. His formative studies of GunaiKurnai culture and practice continue to inform scholarship today.

In 1890, Howitt gave a paper to the Royal Society of Victoria entitled 'The eucalypts of Gippsland' (see also discussion of Howitt's paper in *Proceedings of the Royal Society of Victoria* (Royal Society of Victoria 1891))(Wardlaw 1997). He explained how White settlement had disrupted Aboriginal burning practices, which, in the words of Bill Gammage, let 'undergrowth fill open forest and grass revert to bush' (Gammage 2011: 322; see also Morgan 2013: 17). Gippsland, Howitt told his audience, was more heavily forested than it had been 50 years earlier when the first White settlers arrived (Howitt 1890: 109–113). To the Aboriginal peoples of Gippsland, Howitt observed, 'we owe more than is generally surmised for having unintentionally prepared it [Country], by their annual burnings, for our occupation' (Howitt 1890: 111).

Howitt's paper had little to say on how the GunaiKurnai burned. Rather, it was significant for stating that they had; it recognised that the GunaiKurnai and their neighbours had used fire to create the landscape that so appealed to Angus McMillan, de Strzelecki and the other 'discoverers' of Gippsland. Among the White settlers of Gippsland were those who knew through observation or anecdote that the GunaiKurnai had fired the land, but this was different: here was a scientist explaining how pervasive and important those burning practices had been, and the enormous effects that flowed from interrupting them. Howitt understood that the GunaiKurnai had never been passive occupants of Country, as was often the conclusion of White settlers (Dow 2004: 67–68).

Howitt had been thinking about these themes for some time. In 1869 a story entitled 'Ella's Dream of How the Trees of Nuntin Forest Died' appeared in the *Gippsland Times* (Anonymous 1869: 4). The author of the story was identified only as a 'contributor'. Carol Dow writes: 'The Nuntin Forest was a belt of dead timber extending from the Lakes inland for about seventy miles. By 1869 the trees "young and old" were dead and dry, and no one knew "the cause of their destruction". Eight-year-old Ella F., a squatter's daughter, fell asleep looking at the forest of dead trees in the moonlight. She dreamt of the trees and related her dreaming to her mother' (Dow 2004: 62). Ella's dream described how the poisoning of a group of Aboriginal people and the misery this wrought on the GunaiKurnai had led to the death of the forest. The story was radical in recognising the iniquities that White settlers had inflicted on the GunaiKurnai and their Country. Dow suggests that it may have been written by Howitt, who at the time was police magistrate in Bairnsdale (Dow 2004: 64). She notes that the writer understood the depth of the GunaiKurnai's connections to Country—the dream was used as a vehicle to make this point—and that 'Howitt later described the tree decline in the Nuntin area and concluded that it resulted from changed fire regimes' (Dow 2004: 62, note 92).

Howitt's work on Aboriginal burning practices in Gippsland and beyond was noticed in the press (e.g., Anonymous 1891), and by some members of the public. In 1914, E.H. Lees of Fairhaven, Mallacoota, wrote a letter to the editor of *The Australasian* newspaper (Lees 1914a). Edward Lees was a surveyor who worked widely with the Victorian government in East Gippsland. He had a keen interest in Aboriginal culture and Australian flora (e.g., Lees 1914b, 1915). In his letter Lees cited Howitt and his own observations to confirm Howitt's position on GunaiKurnai burning practices and their interruption by White settlers. Lees (1914a: 8) added: 'Bush fires have long preceded settlement, and ages before the advent of the white man vast areas were burnt off by the aboriginal occupiers accidentally and intentionally when hunting game'. The 'best friend to forest and settler', Lees wrote, 'is moderate firing!'.

It is thought that Lees, who died on 30 June 1921, lived in East Gippsland for much of his adult life. The Victorian Public Record Office in North Melbourne holds extensive records of his work as a surveyor. A selection of these primary documents was consulted for this chapter. While this sample did not include information about his interactions with GunaiKurnai Country and burning practices, other records might. Further research into Lees and his career would be worthwhile.

FORGETTING THE PAST

References to GunaiKurnai burning practices in the 19th century written record range from vague and largely ignorant mentions to the well-informed comments by Howitt. That this record is relatively rich owes something to 19th century settlers in Gippsland having need of fire. Whether or not they understood the science of GunaiKurnai burning, and it seems that most did not, they did know that fire could help them: it cleared land and produced green pick for stock.

This recognition is less evident in the 20th century literary record, perhaps because some settlers saw less need for fire: fertile land had been cleared, farming technology had advanced, and life on the colonial frontier in Gippsland was ever less precarious. By the 20th century there was also, very often, a greater distance between White settlers and GunaiKurnai. The first White settlers in Gippsland had a closer view of GunaiKurnai culture and practice than did their descendants. Many descendants might have inherited a fear of fire, but at the same time forgot what their forebears had learned, one way or another, from the GunaiKurnai: that fire is a useful and important tool.

Those beyond Gippsland also wrote less about the ways in which GunaiKurnai tended Country. In 1966, E.C.F. Bird, a geographer at the University of Melbourne, wrote of the Gippsland Lakes before White settlement: 'the region was occupied only by a sparse aboriginal population, whose activities left little mark on the landscape ... The effects of a nomadic population of hunters and gatherers of food on the vegetation and fauna of this region cannot be determined, but they were undoubtedly trivial compared with the impact of settlers in the last 120 years, and for practical purposes the Gippsland Lakes may be regarded as "unmodified by man" at the time of McMillan's discovery' (Bird 1966: 56).

As non-Aboriginal people began to better understand the significance and sophistication of Aboriginal care for Country, observers wrote of GunaiKurnai burning practices in increasingly nuanced ways (see Griffiths and Russell 2018). In 1969, the archaeologist Rhys Jones published his influential article 'Firestick farming' (Jones 1969: 224–228), and in 1972 the historian Keith Hancock wrote his pioneering book *Discovering Monaro*. Hancock showed how Aboriginal people had managed the land, and in an area not far from GunaiKurnai country. The book's sub-title was *A Study of Man's Impact on his Environment*.

Others took the understanding articulated by Jones and Hancock and applied it to Gippsland. In a 1985 report entitled *A History of the Aboriginal People of East Gippsland*, Kym Thompson wrote sensitively about GunaiKurnai culture and practice. The report notes that 'Aborigines were actively modifying their environment, most particularly by using regular widespread burning to maintain advantageous non-climax vegetation patterns' (Thompson 1985: 52). Non-Aboriginal people have come to know some of the things that Aboriginal people knew all along. It is now understood that the GunaiKurnai fired the land as part of caring for Country, and that they did this with great knowledge and expertise. They knew when and what to burn, and how hot to set the fire.

Typically the literature on Gippsland written in the latter half of the 20th century rarely mentioned the GunaiKurnai. When it did, the reference tended to be cursory, and sometimes derogatory (e.g., Spurrell 1976: 1). In such histories, often little more than a catalogue of 'White man's progress', there was no room for discussion of GunaiKurnai culture, let alone burning practices. Other histories mentioned GunaiKurnai culture, but as a dead relic of a distant past. A history kit prepared for Bairnsdale school students in 1981 included information lamenting the destruction of the GunaiKurnai, now 'gone forever' (Douglas 1981: no page numbers).

ORAL HISTORIES

It remains that the written records mentioning GunaiKurnai burning practices in the 20th century are curiously thin. This literature review is not exhaustive, and no doubt there are worthwhile sources not captured in this chapter. Memoirs, including unpublished memoirs held by Gippsland families, might prove particularly important sources. An even more promising way to strengthen historical knowledge of GunaiKurnai burning practices may lie in oral history. It seems that much discussion of GunaiKurnai burning, especially in the 20th century, never made it to the printed page.

Gippsland was one part of Victoria affected by the January 1939 'Black Friday' wildfires. In his landmark Royal Commission report into the fires, Leonard Stretton (1939) wrote of graziers' knowledge of fire, and of an oral tradition. The following passage is from the environmental historian Tom Griffiths' *Gippsland Heritage Journal* article about Stretton and his report:

> In the drier forests (but generally not the wet mountain ash forests which had less grass), graziers used fire as Aboriginal people had done: to keep the forest open, to clean up the scrub, to encourage a 'green pick', and to protect themselves and their stock from wildfire. In autumn a portion of each run was burnt. It was a tradition handed down over generations, sanctioned, as Stretton observed, by long usage. It was this habit of burning that generated increasing government opposition to the cattlemen and women. (Griffiths 2002: 10).

Griffiths adds: 'The Royal Commission was nothing less than a full-scale enquiry into Australian bush culture. The language the settlers and farmers used—"burning to clean up the country"—was uncannily like that of Aboriginal people' (Griffiths 2002: 13).

When in the 1830s James Macfarlane drove cattle from the Monaro into what is now Victoria, he followed an ancient Aboriginal pathway. Both the Aboriginal inhabitants of the area, and Macfarlane and other cattlemen who used this path, had an interest in keeping the country clear. And they did so up until recent decades. GunaiKurnai Elder Russell Mullett, speaking of the second half of the 20th century, told me that cattlemen using Macfarlane's Track would defer to Aboriginal knowledge and either follow instructions or ask their Aboriginal workers to set fires to keep paths open. In how many places in Gippsland might there be similar stories?

Only in recent decades, with tighter restrictions on the lighting of fires, has the bush around Macfarlane's Track closed in (Figure 3.2). Bill Gammage, who has written extensively about Aboriginal burning practices, including in Gippsland, was taken to Macfarlane's Flat by mountain cattlemen, members of families long established in the High Country. They told him about the spread of scrub in recent decades, and of how their forebears had relied on fire to keep country open. Their spoken memories match the knowledge shared by Russell Mullett.

Chris and Jeanette Commins of Ensay North are graziers long-established in the High Country. Their family has lived in and around the High Country since the 1840s, and have run cattle in Gippsland for the past 100 or more years. When James Lilburne Commins, Chris's grandfather, came to Ensay after the First World War, fire was part of Gippsland life. James Arthur Commins, Chris's father, grew up at Ensay in the 1920–1930s. He remembered smoke as ubiquitous and constant, including in summer: the haze came from bushmen burning off and from fires caused

Figure 3.2. Macfarlane's Track and Flat. Traditionally this landscape was cleared regularly using managed fires. A: Macfarlane's Track, near the Murray River, showing overgrown vegetation that has resulted from not burning the landscape. B–D: Macfarlane's Flat, showing edges of grassy flats (photographs: Joanna Fresløv, 15 April 2019).

by lightning. Chris is sure that Aboriginal knowledge of fire has informed the practices of his family and, more widely, the burning methods of mountain cattlemen, with observation and anecdote the channels along which information has passed across cultures and generations. For instance, he remarks that fire should never burn so hot and high as to scorch canopy. Reducing canopy cover allows in light, which promotes 'massive regrowth creating a thicket of scrub and elevated litter' (Chris and Jeanette Commins, personal communication, Ensay North 30 July 2021). The threat of intense crown fires informs how Chris and Jeanette manage land (Figure 3.3), and he takes an active interest in the application of the firestick and cool burning. For him the utility and need for such burning is obvious (Figure 3.4).

In a separate interview, Ewan Waller made exactly the same point: fire should not touch the treetops. Waller and his forebears have farmed at Glenaladale since the 19th century. His grandfather, Tom Morrison, burned for green pick. He did this in January, and burned hot: several fires got away from him, with devastating effects. The bush taught Basil Waller, Ewan's father, to do otherwise. He knew to burn cool. He always carried a box of matches and in winter would drop fire to remove tussocks and blackberries. Very few fires escaped him because he knew how to burn. Ewan wants to see more cool fires in Gippsland and beyond. He uses the words 'gentle' and 'respectful' to describe the role of cool burning in caring for Country (Ewan Waller, personal communication, Bairnsdale 13 August 2021).

Figure 3.3. Photograph of land at Nunniong, taken in November. The difference in the length of the grass is the result of different grazing practices rather than burning. The photograph is included to show how mountain cattlemen and women adopt particular practices to manage country and fire risk. In a fire the patch at left would burn, while the patch at right and beyond the fenced area would not (photograph: Chris Commins c. 1990).

John Mulligan, born in Orbost in 1931, is a fourth-generation member of an East Gippsland settler family. The following quotes are taken from his recollections, published online on 3 February 2020 (Mulligan 2020). 'When my grandmother's older sister (Mrs Coleman) first came to Mallacoota (ahead of the arrival of my grandparents), she said there was a small band of aborigines, who moved about, burning wherever they went'. That was about 1890. When Mulligan was a boy, 'fire was a constant in the bush. Everyone learned to live with it ... Bush dwellers of the time had a completely different understanding of the necessity of regular fire in the environment and its acceptance, than that of the majority of people today'. The effects of the 1939 Black Friday fires in East Gippsland, Mulligan recalls, were not as severe as elsewhere because graziers had been in the habit of burning regularly (Jurskis 2015: 166). They burned for feed and to keep the forest understorey clean.

Margaret Mulligan, John's wife, is another Gippslander who grew up with fire. Her family lived near Yarragon and later at Mallacoota. Margaret's father would tell his daughters to 'Get on your horses and go out and burn today' (Margaret Mulligan, personal communication, Wangaratta, 11 August 2021). He knew when conditions were right, as did they. Margaret says that this knowledge was in-bred; they knew to burn in autumn, where to drop a match, and how to keep fire cool and contained.

Writing in 1952 about White settlers in the forests of the Cann Valley, D.M. Thompson noted 'a tradition of fire handed down over three, and in places four, generations' (Thompson

Figure 3.4. Photograph of well-tended land on the Snowy Plain in the New South Wales High Country. The owner of this property used the firestick and grazing to care for Country. This land, privately held, is within the Kosciuszko National Park. It was not razed in the 2003 wildfires that afflicted the area, unlike much of the surrounding land. When surveying this country in the 1840s, Thomas Townsend, Deputy Surveyor General for New South Wales, wrote: 'The blacks had visited the Snowy Mountains, a short time previously to us, for the purpose of getting "Bogongs", a species of moth, about an inch long, of which they are particularly fond; to obtain them they light large fires, and the consequence was, the country throughout the whole survey was burnt, leaving my bullocks destitute of food. During the time I was on the range the lower parts of the country were burning, and I was prevented, in almost every instance, from getting angles on any distant points, by the dense masses of smoke obscuring the horizon in all directions' (cited in Jurskis 2015: 68–69). (photograph: Chris Commins c. 1990).

1952: 2). Nowhere in his University of Melbourne thesis did he mention the GunaiKurnai. The Mulligans, too, draw no link to Aboriginal knowledge and practice. They see their own knowledge of fire as something inherent among older 'bushies', people who know the land. In contrast, a 2010 paper produced by The Mountain Cattlemen's Association of Victoria did draw an explicit link. It connects the burning methods of graziers to Aboriginal practices, noting that from the 1830s onward mountain cattlemen 'lit cool fires in the autumn after mustering' (The Mountain Cattlemen's Association of Victoria 2015: 3). Vic Jurskis, a forester of long experience with knowledge of Kosciuszko country, states that Aboriginal people, having gathered on the high plains to feast on Bogong moths, would burn the bush on their various ways home to ensure that access was maintained for the next season. He understands that alpine seasonal graziers adopted a similar practice: as they left the High Country in autumn, they burned grass to stimulate the next season's growth, and bush to keep country clear (Vic Jurskis personal communication by Zoom, 23 August 2021).

The oral histories of burning held within non-Aboriginal Gippsland families are worth recording (see Wakefield 1970: 153). They show that living close to the land, as farmers and cattlemen and women do, pushes people to learn something of Aboriginal practices, whether consciously or not. Necessity demands it. It follows that GunaiKurnai burning practices are probably better represented in the non-Aboriginal oral records than they are in the non-

Aboriginal written records: this reliance on a spoken tradition offers a surprising parallel between White and Aboriginal histories of fire. How many cattlemen and other non-Aboriginal Gippslanders have drawn on GunaiKurnai burning practices without having thought to say so? How many burn without recognising a reliance on an older, Aboriginal tradition? Are there areas of Gippsland where GunaiKurnai burning knowledge has been applied continuously since White settlement without the written record taking notice? These and other questions about the transfer of burning knowledge from Aboriginal to non-Aboriginal communities are worthy of further exploration.

In collecting these sorts of histories, there is an example to follow. Daryl Tonkin was a White man who spent much of his life living with and learning from GunaiKurnai people. His partner and the mother of his children was Euphemia Mullett (née Hood), a GunaiKurnai woman. Their property at Jackson's Track near Drouin in West Gippsland became a home, workplace and haven for many GunaiKurnai families. Tonkin knew the bush and he knew fire. Late in life he began to commit some of his life story to paper, a decision that led to the celebrated book *Jackson's Track* (Tonkin and Landon 1999: e.g., 208). Until he began to share his memories, what he had learned of Aboriginal culture and practice remained hidden from all but his family. Tonkin was a remarkable and deeply unusual man, and it would be naïve to think that such accounts lie hidden around every corner in Gippsland. Equally, it would be foolish to think that his was the last word. There must be other such accounts, perhaps not as rich, but important nonetheless, to be added to the historical and written records of Aboriginal burning practices in Gippsland and the country beyond.

Chapter 4

Eugene von Guérard on GunaiKurnai Country 1860–1861: Reading the Story of Fire in his Depictions of the Landscape

Ruth Pullin

INTRODUCTION: THE JUNCTION OF THE BUCHAN AND SNOWY RIVERS

On 9 January 1861, Eugene von Guérard (1811–1901) sketched the magnificent view of the Snowy River at the point where, winding through hilly, forested country, it is joined by the Buchan River (Figure 4.1). This landscape was shaped by fire. The evidence is embedded in the fabric of the landscape as recorded by the artist—in the park-like openness of the woodland, the limited undergrowth, the distribution of the trees and the areas of open grass. Evidence of 'templates', the term used by Bill Gammage (2011) to identify the inter-generational patterns of land management that involved the use of controlled burns, can be found throughout von Guérard's drawing. He saw and documented the open corridors of grass, framed by tree belts, that run down the steep east-facing slopes to the Snowy River. The grass on these corridors, the fresh pick generated by the application of cool burns, was designed to attract grass-eating mammals and to 'clean Country'—to keep it healthy—as we now know from Aboriginal knowledge-holders (see Chapters 1 and 3). From the shelter provided by the tree belts, hunters could corral their prey down the steep corridors to the water. These grass corridors were, in effect, traps. The mosaics or patches of open grassland on the otherwise forested crest and west-facing slope of the ridge on the left of von Guérard's drawing, and on the hills in the middle distance, are also the result of cool burns; these grass patches, like the corridors, were designed to attract and ambush prey. An example of this was recorded for GunaiKurnai Country by the Church of England missionary the Reverend John Bulmer, who managed the Lake Tyers Mission Station at Bung Yarnda from 1862 to 1907. Bulmer recorded that '... in summer they set fire to a large tract of country and speared the animals as they were escaping from the fire. They also got many after the fire almost roasted enough for eating' (cited in Vanderwal 1994: 61). In the absence of evidence for changes in soil, microclimate or other conditions, Gammage (2011: 8–9, 21–95) suggests that patch-burning is the only viable explanation for the areas of open grass located in bushland observed by early colonists and colonial artists. The locations of such patches were changed over time to ensure that animals did not become wary of particular places, while also allowing for the bush to recover. The critical determinant for their location was the association of water, grass and forest (Gammage 2011: 61). In von Guérard's drawing, the grassy corridors lead to water, grass mosaics are on forested hills close to the river, and hills further from the water are densely vegetated.

Figure 4.1. Eugene von Guérard, Junction of the Buchan River with the Snowy River. 9 January 1861 (1861), pencil and ink (from 'Australian Sketches 1860–1861', Alexander Turnbull Library, Wellington, New Zealand, E-337-f-034).

VON GUÉRARD'S DRAWINGS AS RELIABLE RECORDS

While von Guérard's associate Robert Brough Smyth was aware by 1878 that it was customary practice 'to burn off the old grass and leaves and fallen branches in the forest, so as to allow of a free growth of young grass for the mammals that feed on grass' (Brough Smyth 1878: xxxiii), it is unlikely that in 1860 von Guérard understood that the grass necks and mosaics that he observed in GunaiKurnai Country were the result of fire practices followed by the Krauatungalung (a GunaiKurnai clan) for thousands upon thousands of years. However, we can be sure that as he drew, working directly from nature, he recorded exactly what he saw. These drawings, many of them elaborated with detailed notes, were his primary references, the documents he relied on for the paintings executed later in his Melbourne studio. For him, topographical accuracy was essential, and his success can be measured in a comparison between his drawing, *Junction of the Buchan River and the Snowy River*, and a photograph taken 160 years later from the same or very close vantage point (Figure 4.2). Descriptive notes refer to the 'dark sheoaks' and 'old wattle' that he saw there on the day, revealing his interest in botanical accuracy, and the reference to the 'white foam' on the water caused by turbulence at the meeting of the two rivers, his attention to detail. Von Guérard would annotate many of his drawings, often in old German, with precise details about features of the scenes he depicted, such as the light and shadows, the colours of the foliage and grassy patches, and the types of trees present. These details, which informed his studio works—oil paintings, lithographs and presentation drawings—also often appear to have been recorded for their own sake; they reflect the artist's innate curiosity and interest in the world he saw.

Figure 4.2. Junction of the Buchan River with the Snowy River, photographed in 2019 a few months before the 2019–2020 Gippsland wildfires (photograph: Bruno David).

All the sketches and drawings discussed in this chapter were drawn by von Guérard on the spot, directly in front of the subject portrayed. These on-site drawings were working documents, made for the artist's own reference. He depicted what he saw without making pictorial or compositional changes to the scene—such alterations were simply not required. Slight variations to a composition were often made in the studio, when the subject was transferred to canvas or prepared for printing as a lithograph, but the overall aim was to remain as faithful to the subject as possible, and to convey its essential 'truth'. Sometimes the features of a landscape were compressed slightly to fit the proportions of a canvas or the lithographic sheet or the mountain peaks slightly heightened for compositional reasons or dramatic effect. The most frequent changes appear in the foregrounds of the studio paintings, where a tree, a fallen log or a group of rocks may be introduced to frame a composition or to accentuate the foreground in order to enhance, by contrast, the illusion of pictorial depth. Von Guérard selected his drawing of the junction of the Buchan and Snowy rivers from 'the hundreds of drawings suitable for publication' for his album of 24 lithographs, *Eugène von Guérard's Australian Landscapes* (1866–1868) (von Guérard 1870) (Figure 4.3). The drawing, at 32.3 × 52.8cm including a narrow margin, is slightly wider than the lithographic image with its squarer dimensions of 33.0 × 50.5cm. Here, as in many other works, von Guérard made subtle and almost imperceptible incremental shifts in, for example, the steepness of slopes, across the whole composition such that the topographical integrity of the view was not compromised.

Figure 4.3. Eugene von Guérard, Junction of the Buchan and Snowy rivers, Gippsland, Victoria (1867), colour lithograph. Plate 9 in *Eugène von Guérard's Australian Landscapes* (1866–1868), published by Hamel and Ferguson, Melbourne (National Gallery of Victoria, Melbourne).

THE ORIGINS OF VON GUÉRARD'S 'TRUTH TO NATURE'

'My wish was, even if not to create a complete work of art, then at least to put before the public views from this part of the world that demonstrate the character of the Australian landscape faithfully and with truth to nature' (von Guérard 1870, cited in Pullin 2011: 25). Von Guérard's passionate interest in nature, and his ability to depict it with precision and in minute detail, can be traced back to the early influence of his artist-father, Bernard von Guérard. At the time of Eugen's birth (he was christened 'Johann Joseph Eugen von Guérard' and known as 'Eugen'), in 1811, his father Bernard was a well-regarded painter of miniature portraits in Vienna. His subjects included high-ranking military men and royalty and aristocrats attached to the courts of Vienna and (later) Italy. In 1827, Bernard and his 16-year-old son Eugen set out on an open-ended sketching and painting trip to Italy. As father and son worked side-by-side, sketching in the Italian landscape, the young von Guérard learnt the skills of a miniaturist, honing his eye for detail, and becoming adept at capturing the minutiae of nature using hard, finely sharpened pencils, pen and ink and, in the studio, the smallest and finest brushes. Von Guérard's commitment to the faithful portrayal of nature developed further when, in Germany in the 1840s, he studied landscape painting under Johann Wilhelm Schirmer (1807–1863) at the highly regarded Düsseldorf Academy. Schirmer encouraged his students to paint nature with an 'obedient, natural sense, that everything seen should be seen as it is, always with open eyes and a warm heart' (Schirmer 1833, cited in Eggerath 2003: 63; see also Pullin 2011: 78).

HUMBOLDT, VON GUÉRARD AND THE MEETING OF SCIENCE AND ART

Von Guérard's view of the natural world was profoundly shaped by the ideas of the brilliant German polymath, traveller and natural scientist, Alexander von Humboldt (1769–1859). The celebrated Humboldt had fired the imaginations of a generation, inspiring, among others, Charles Darwin, with his ground-breaking discovery of the 'interconnectedness' of all aspects of the physical world. The impact of his ideas was felt around the world. Von Guérard was one of a cohort of eminent German artists and scientists, including the botanist, Ferdinand von Mueller (1825–1896) and the geophysicist, Georg von Neumayer (1826–1909), whose Australian careers were informed by Humboldt and his work (Pullin 2011).

Humboldt devoted an entire chapter of the second volume of his best-selling publication, *Cosmos: A Sketch of a Physical Description of the Universe* (1847), to the subject of landscape painting. He urged landscape painters to 'seize' on 'the true image of the varied forms of nature', to depict individual plant species with scientific accuracy, not introduced from afar into hothouses but in the context of their natural growing environments (Humboldt 1847: 452; 1849: 229). Along with other influential figures in Germany, notably the scientist, theorist and landscape painter Carl Gustav Carus (1789–1869), Humboldt argued that art and science were parallel and complementary disciplines, each capable of informing the other (Pullin 2011: 15). Together artists and scientists could provide complementary information about the natural world, its geomorphology and geological processes, soil and climatic conditions, plant species and their distribution and relationships in relation to latitude and elevation above sea level, and more. He saw that in one image the artist could communicate all this with an immediacy unavailable to the scientist, who may require thousands of words to convey the same information.

In Australia, and in keeping with the empirical methodology of Humboldtian science, the precise observation of landscapes, many of which were as yet unseen by European eyes, was fundamental to von Guérard's vocation as an artist. 'Descriptions of nature', Humboldt argued, 'may be defined with sufficient sharpness and scientific accuracy, without on that account being deprived of the vivifying breath of the imagination' (Humboldt 1847: 438). Von Guérard's drawings are alive with his wonder at the natural world as he experienced it on his sketching expeditions across southeastern Australia.

TRAVELLING WITH SCIENTISTS IN GUNAIKURNAI COUNTRY: HOWITT AND NEUMAYER

Humboldt's ideas about art and science were realized in practice when von Guérard travelled alongside scientists and scientific men on their research and exploratory expeditions. He spent three of his seven weeks in GunaiKurnai Country in December 1860–January 1861 travelling with his friend, the experienced bushman, explorer, natural scientist and ethnographer Alfred Howitt (1830–1908) (Pullin 2018: 214–221). Joining Howitt and his party on their Government-sponsored gold-prospecting expedition into the rugged Gippsland Alps gave von Guérard access to country he could not have penetrated alone (Figure 4.4). The scientifically oriented Howitt went on to make significant contributions to Australian botany—his *Eucalypts of Gippsland* was published in 1890—geology and anthropology. For von Guérard, this trip was an opportunity to see and discuss the landscape with an informed friend, a friend he knew well having travelled with him on an expedition through the rugged Yarra Ranges to Mount Baw Baw in 1858. The richness of their exchange is nowhere more evident than in von Guérard's large pencil drawing, later reworked in ink, of the Moroka River gorge (Figure

4.5). It is a drawing of extraordinary detail and complexity, geological precision and power. Howitt, recounting their experiences on that day, described being 'suddenly brought up by a precipice of three hundred feet which runs sheer into the river—which just beyond shoots over a fall of perhaps 30 feet'. While the men are attempting to navigate a way forward, one of them 'looking very much like a fly crawling up a house side' as he clambers up a precipice, 'De Guérard sits down to sketch the rocks—I sit with him and light my pipe' (Howitt 1860). Late in 1862, this time travelling with another good friend, the Humboldtian scientist Georg von Neumayer and his party on a magnetic survey of northeast Victoria, von Guérard again passed through GunaiKurnai Country when the party skirted through the northern-most reaches of Brabralung clan's Country on their approach to Mt Kosciuszko from the south.

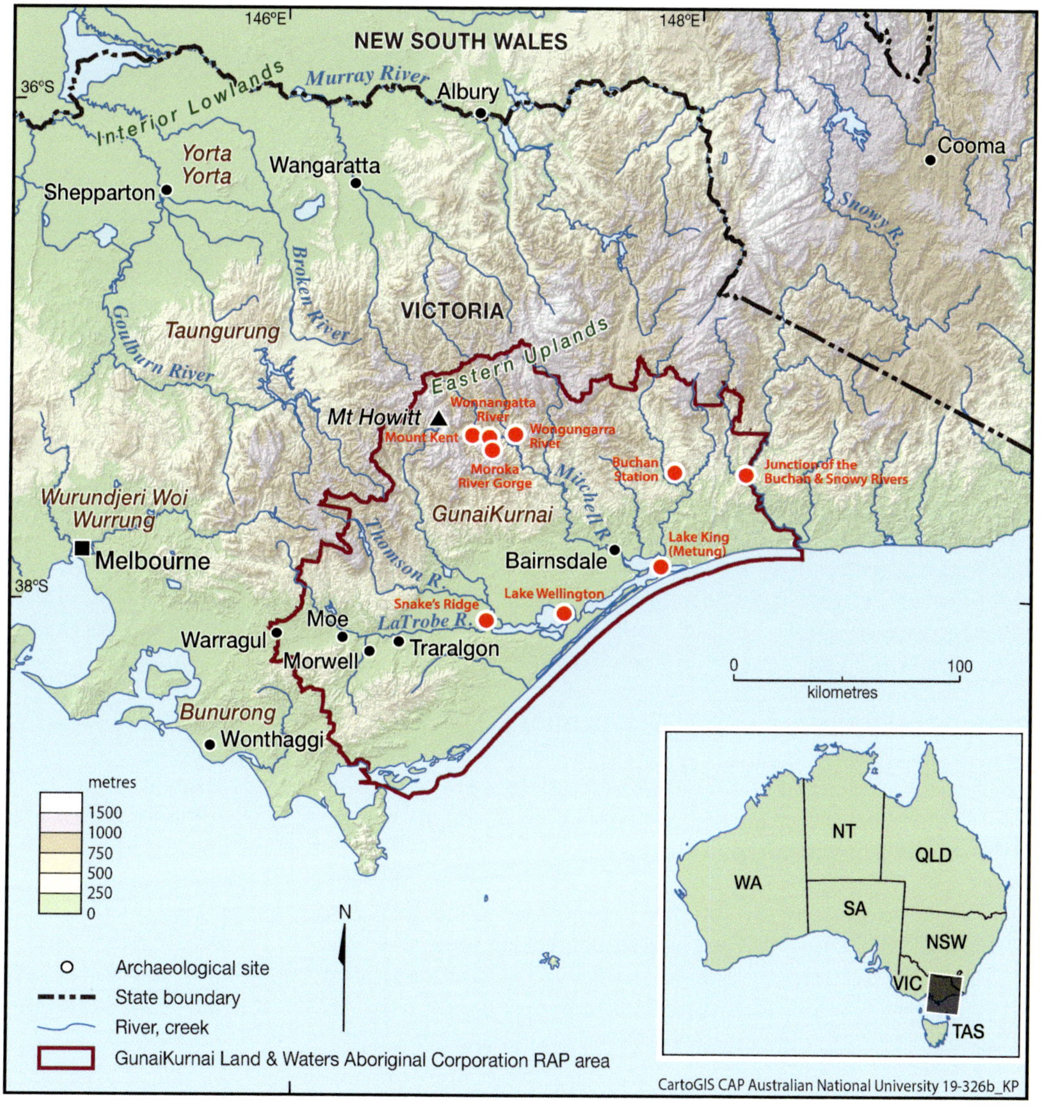

Figure 4.4. Places where Eugene von Guérard drew landscapes in GunaiKurnai Country, as discussed in this chapter.

Figure 4.5. Eugene von Guérard, Moroka River near Mt Kent Gippsland. 13. Dec. 1860 (1860), pencil and ink (from 'Australian Sketches 1860–1861'. Alexander Turnbull Library, Wellington, New Zealand, E-337-f-017).

On the 1860–1861 Gippsland expedition, von Guérard filled all but the last four pages of his pocket-sized sketchbook with pencil sketches of GunaiKurnai Country, and he produced 35 larger drawings, most later finished in pen and ink, all now held in a bound album in the Alexander Turnbull Library in Wellington, New Zealand. The number and beauty of the drawings inspired by this trip accords with Howitt's observation that the artist was in a 'state of delight' as, between 1–20 December 1860, they trekked over and through the valleys of the Wonnangatta, Wongungarra and Crooked rivers, over the Snowy Bluff, into the Moroka Gorge and back to their base camp on the Wonnangatta River (Howitt 1860).

RECORDS OF CULTURAL BURNS IN THE HIGH COUNTRY

Bill Gammage identified von Guérard's *View of the snowy bluff on the Wonnangatta River, Gippsland Alps, Victoria* (1864) (Figure 4.6), as an example of the artist's work in which evidence of fire patterns, in this case patches and mosaics, can be seen:

> At centre and right von Guérard shows three sloping clearings split by tree-filled gullies. They face northeast to catch the sun and bring animals to feed and warm. On them patch-burns located animals, and let hunters drive them uphill or headlong into a gully. Two clearings also carry lone trees spared by frequent grass fires, even when young. Perhaps rocks or backburning protected them. (Gammage 2011: 68).

The presence of areas of 'beautiful meadowland' (Gammage 2011: 7), on otherwise forested land where no shift in aspect or soil conditions was evident, had baffled early European travellers in other parts of the country. What von Guérard understood as to the reasons for

Figure 4.6. Eugene von Guérard, View of the snowy bluff on the Wonnangatta River, Gippsland Alps, Victoria (1864), oil on canvas, 95.2 × 152.7cm (National Gallery of Victoria, Melbourne).

the existence of such grassy areas, we cannot know, but we can be sure that he saw them. He hinted at their existence in a double page pencil drawing of this view in his sketchbook (Figure 4.7) but a larger, more detailed drawing, if it existed, has not survived.

Figure 4.7. Eugene von Guérard, Snowy Bluff 19 Dec. 1860 (1860). Sketchbook XXXII, No. 13–14 Australian, Dixson Galleries, State Library of New South Wales, Sydney, DGB16, v. 11, f. 31.

Figure 4.8. Eugene von Guérard, Mt Kent, Gippsland 20 Dec. 1860 (1860), pencil and ink (from 'Australian Sketches 1860–1861', Alexander Turnbull Library, Wellington, New Zealand, E-337-f-022).

Comparable mosaics of open grassland are clearly visible in a highly finished pencil, ink and wash drawing, *Mt Kent, Gippsland* (20 December 1860), drawn on the day after his sketch of the Snowy Bluff (Figure 4.8). The two grassy patches, on the sunny gentle lower slopes below, and to the left of, Mount Kent, are described by von Guérard in his notes as 'gelbe Wiesen', yellow grasslands. They are identical to those that Gammage described as the result of patch-burns in von Guérard's *View of the snowy bluff on the Wonnangatta River*.

When von Guérard decided to make Mount Kent a subject of an oil painting, he opted for the view recorded in a very cursory pencil drawing in his sketchbook, rather than the more detailed drawing discussed above (Figure 4.9).

The open foreground of the painting, *Mount Kent, on the Wonnangatta, Gipps Land* (1873), like the sketchbook drawing, is populated with a few isolated eucalypts and a dead tree (Figure 4.10). They have grown tall and straight, reaching for the light, suggesting that as they grew they were surrounded by other trees. Now, as solitary individuals exposed to the light, epicormic branches (growths that sprout from dormant buds along a trunk or branch) have sprouted along the length of their trunks, and a handful of shrubs have sprung up on the grassy rise along which a mountain creek flows and where emus graze. Cool fires do not cause scrub to germinate or impact trees in the ways pictured here, suggesting that this area has been affected by a hotter fire, perhaps one caused by a natural event like a lightning strike, or European actions, or by another cause? Research has shown that 'high country fire frequency' increased following the arrival of Europeans (Gammage 2011: 172, citing Banks 1997: 9–12).

Figure 4.9. Eugene von Guérard, Mt Kent and Part of the Snowy Bluff 1860 (1860). Sketchbook XXXII, No. 13–14 Australian, Dixson Galleries, State Library of New South Wales, Sydney, DGB16, v. 11, f. 32.

Figure 4.10. Eugene von Guérard, Mount Kent, on the Wonnangatta, Gipps Land (1873), oil on academy board, 31.0 × 47.0cm (private collection).

Figure 4.11. Eugene von Guérard, Wonangatta [Wonnangatta] River below the Junction of the Moroka River, Thursday 11 Dec. 60 (1860), pencil and ink (from 'Australian Sketches 1860–1861', Alexander Turnbull Library, Wellington, New Zealand, E-337-f-015).

The steep slopes of the mountains depicted by von Guérard in his paintings and drawings are invariably densely vegetated, unlike the ridge-tops leading to the High Country and elevated plains, which are more open. This accords well with GunaiKurnai Elder Russell Mullett's knowledge of Country, passed down the generations, where the Old People, the ancestors of today's generations, 'travelled mainly along the ridgelines up into the High Country'. These 'routes from the foothills to the mountains were marked by cosmological and other cultural associations', and were thus maintained as 'healthy Country' including through cool patchwork fires by the GunaiKurnai (Fresløv and Mullett, in press). With further study it may be possible to argue conclusively that von Guérard's drawings show evidence of cool patchwork fires in the High Country. In one, *Wonangatta* [Wonnangatta] *River below the Junction of the Moroka River* (11 December 1860), what appears to be a series of open, grassy belts lined with trees can be seen on the slope, leading down to the river, and facing the viewer (Figure 4.11). While it may be that a site visit will reveal other explanations for these open, treeless corridors—perhaps a sequence of rock ridges—further investigation is warranted.

EVIDENCE OF CULTURAL BURNS ON APPROPRIATED GUNAIKURNAI LAND

Howitt and von Guérard may have thought, as suggested by an art critic for the *Argus* in a review of von Guérard's 1876 version of the Snowy Bluff composition, 'that they were the first human beings to penetrate into its [the Wonnangatta Valley] sequestered solitudes, as there was no trace of even a black man's presence' (Smith 1876: 5). By contrast, the imprint of European intervention was clearly evident in the GunaiKurnai country on which von Guérard

Figure 4.12. Eugene von Guérard, From Mr John King's Snake's Ridge, Gippsland, 19 and 10 November, 1860 (1860), pencil on paper (from 'Australian Sketches 1861', Alexander Turnbull Library, Wellington, New Zealand, E-337-004).

travelled before and after his three weeks with Howitt in the Gippsland Alps, as were signs of the land management practices of its Traditional Owners. On his way to join Howitt, von Guérard had spent time on John King's station, Snake's Ridge, and Angus McMillan's Bushy Park. Both men commissioned paintings of the properties they now regarded as their own. Such paintings were a kind of affirmation of a property owner's legitimate claim to the land—a pertinent issue in the context of the *Land Sales Act* of 1860. Property portraits such as these could confer a level of respectability and status to the landowner, veiling over the realities of land theft and violence. King acquired the rights to the extensive Snake's Ridge run, which he had managed for some years, in 1851 (Figure 4.12).

Rather than his homestead, it is his achievements as a pastoralist and a cultivator of the land that are celebrated in von Guérard's *Mr John King's Station, Gippsland* 1861 (Figure 4.13). However, it was not King who was responsible for the lush grasslands on which his cattle grew fat; they had been nurtured by GunaiKurnai over thousands and thousands of years. This was not a 'natural' ecosystem after all, but one which had been managed. Evidence of the judicious use of fire (the 'firestick farming' of Rhys Jones (1969)) to encourage the fertile grasslands and for hunting was visible when von Guérard recorded the scene in 1860. Like the 'sawtooth tongues of forest' which 'bite into grassland to let hunters ambush prey,' which Gammage saw in von Guérard's lithograph, *The Sources of the River Wannon* (1866), here promontories of forest extend in bands over the grassland between King's station and the mountains (Gammage 2011: 59). This common template was recorded by von Guérard throughout western and northeastern Victoria.

Figure 4.13. Eugene von Guérard Mr John King's Station, Gippsland 1861, oil on canvas on board, 40.7 × 83.9 cm (private collection).

After he left Howitt on 23 December 1860, von Guérard continued south along the Avon River to Lake Wellington and then east towards Buchan and the Snowy River, staying at and sketching properties on the way, including Bushy Park, where he sat with GunaiKurnai men and watched as, in the space of one and half hours, they made a canoe; Mr Bolden's Station, Strathfieldsaye; and Mr Smith's station, Lindenow. At Strathfieldsaye he was captivated by the beauty of Lake Wellington, and the abundance and diversity of the shrubs and grasses on its shoreline—including Marley Point, part of a dune system which GunaiKurnai Elder Russell Mullett (personal communication February 2021) points out was a burial ground for the people of this area (Figure 4.14). In his sketchbook study of Lake King, the artist recorded the series of silt jetties known by the GunaiKurnai as Wandin (boomerangs). These and other drawings hint at the significance von Guérard's drawings may hold for the future as records of GunaiKurnai Country, its landforms, lakes, rivers and vegetation, as they were in 1860–1861 (Figures 4.15, 4.16).

Figure 4.14. Eugene von Guérard, West of Lake Wellington Rosenith [Roseneath] 2 Janner (sic) 1861 (1861), pencil and ink. Sketchbook XXXII, No. 13–14 Australia, 'Gippsland 1860'. Dixson Galleries, State Library of New South Wales, Sydney, DGB16, vol. 11, f. 41. Left: Strathfieldsaye. Station v. Mr Bolden. 2 Jan 61 Lake Wellington. Right: West of Lake Wellington Rosenith [Roseneath] 2 Jan 61.

Figure 4.15. Eugene von Guérard, Mountains N.E. of Lake King, Gippsland January (1861), pencil. Sketchbook XXXII, No. 13–14 Australia, 'Gippsland 1860'. Dixson Galleries, State Library of New South Wales, Sydney, DGB16, vol. 11, f. 46.

Figure 4.16. Eugene von Guérard, Point Metung & Exit of the Lakes to the Sea. Saturday 12 January 61 (1861), pencil and ink (from 'Australian Sketches 1860–1861', Alexander Turnbull Library, Wellington, New Zealand, E-337-f-036).

ON THE WAY TO THE JUNCTION OF THE BUCHAN AND SNOWY RIVERS

On the night before he reached the junction of the Buchan and Snowy rivers, von Guérard stayed with John MacLeod at Buchan Station. While in the area he sketched a view of Mt MacLeod—misspelt on his sketch as 'Mac Claude'—in which three mosaics of grassland are clearly visible on the forested slopes that he described as bathed in 'soft morning light from the east' (Figure 4.17). Did he, during his time at MacLeod's station, take the opportunity to talk with the Brabralung people he met there, as he had with GunaiKurnai men at Bushy Park? By 1861, MacLeod was employing Brabralung men as stock riders, and his sister, reportedly, 'taught local Brabralung women how to sew'. His property was thought to be 'a good place for an Aboriginal reserve', and MacLeod's appointment as Honorary Correspondent of the Central Board for watching over the interests of the Aborigines was imminent (Howitt and Fison no date). Tulaba (Billy Macleod), a Brabralung man, was to be Howitt's main informant, and worked as a stockman for the Mitchell and Snowy Rivers squatting stations near Buchan, Orbost and Bairnsdale (Mialanes *et al.* in press). In his large and detailed drawing of MacLeod's station on the Buchan River, von Guérard portrayed a group of the people then living there, seated in front a bark shelter close to the main house (Figure 4.18).

The next day, von Guérard followed the Buchan River to its junction with the Snowy River, where he quickly identified the optimum vantage point for his dynamic composition (see Figure 4.1). From an elevated position looking northeast, the Snowy River reads as a powerful arc as it sweeps around a promontory (described by von Guérard as a 'half island') to be joined

Figure 4.17. Eugene von Guérard Mt Mac Claude [MacLeod] 8 January 61 (1861), pencil and ink. Sketchbook XXXII, No. 13–14 Australia, 'Gippsland 1860'. Dixson Galleries, State Library of New South Wales, Sydney, DGB16, vol. 11, f. 47.

Figure 4.18. Eugene von Guérard, Buchan Station & Mt Dawson, 8 & 9 January 1861 (1861), pencil and ink (from 'Australian Sketches 1860–1861', Alexander Turnbull Library, Wellington, New Zealand, E-337-f-033).

by the Buchan River. The meticulous detail and topographical accuracy of his drawing and the related lithograph are remarkable documents, illustrating the precise condition of the country as it was on 9 January 1861, at a time of year that is today the peak of the wildfire season. Plant species, vegetation density and distribution, the relative openness of the woodlands, and the patches and necks of grassland flanked by forest, all indicative of the use of fire to manage the land, have been reliably recorded. The value of his record of the condition of this landscape, as it was in 1861, was brought into sharp focus by the wildfires that swept through this area at the end of 2019 and early 2020. At that time, Donald Graham (2020) who, with his wife Bronwyn, lived above the junction of the Buchan and Snowy rivers at the precise location where von Guérard had drawn this scene, observed of his lithograph and the physical landscape:

> He painted what was there and not what he thought should be there. And from that you can see what the landscape was like then ...
>
> Over the last 150 years it's changed from an open woodland because of the different practices or lack of burning if you like—a whole multitude of different approaches. It's become forest and that ranges from pockets of rainforest to woodlands. But worse than that, it's a forest choked with an understory of bark, dead leaves, debris and fallen trees to the point you couldn't walk through it. (See Figure 4.2).

In 1870, von Guérard wrote of his 'desire to imitate nature so well as in his power', convinced that 'for the future his paintings would have greater value' (von Guérard 1870). With the increasing awareness of the environmental significance of his work, and perhaps without von Guérard being fully aware of how the 'natural' vistas he observed had in fact been created by countless generations of GunaiKurnai managers of Country, it seems he was right.

Chapter 5

20th and 21st Century Wildfires and Prescribed Burning in GunaiKurnai Country

Jessie Buettel, Bruno David and Stefania Ondei

In order to understand the impacts of wildfires and prescribed burning on cultural sites in GunaiKurnai Country, it is important to establish where and when these fire events have occurred (i.e., the pattern of fire history). For this chapter, we have used data on fire history and the spatial extent of wildfires and planned burns ('prescribed burning') recorded since 1903 by the State of Victoria's Department of Environment, Land, Water and Planning (DELWP). The data are freely available through the Victorian Data Portal at https://discover.data.vic.gov.au/dataset/fire-history-records-of-fires-primarily-on-public-land. These data consist of layers of geographical information—in the form of spatially-explicit shapes/polygons—that outline where, when, and to what extent wildfires and prescribed burns have occurred across the landscape. In this chapter, we use Geographical Information System (GIS) mapping to overlay these spatial layers across the GunaiKurnai RAP area to explore and discover fire patterns across space and through time. This allows us to ask questions such as: Have wildfires increased in frequency since the 19th century? How large were the wildfires, and when did they occur? Which features of the landscape (e.g., vegetation type or human footprint) might facilitate higher wildfire and prescribed burning frequency and extent? Answering such questions and exploring temporal and spatial patterns of landscape fires is a critical step for determining the past, current and future impacts of fire on cultural sites.

Although the fire history spatial layer from DELWP represents the spatial extent of fires from 1903 across Victoria, there were no records of fire events in GunaiKurnai Country for the years 1903 to 1927 (see Chapter 15 for comments on the robustness of the early spatial data). Therefore, in this chapter and throughout this monograph, we quantify patterns and display fire information from 1930 to the 2019–2020 fire season. It is also important to note that the fire history displayed in Victoria's spatial layers also includes information on Traditional Owner fires (often referred to as 'cultural burns'). Nineteen fires of this type have been recorded across Victoria, the oldest in September 2019. More than half (10) of the Traditional Owner fires that were reported were less than 1ha in extent, the remaining nine being less than 5ha each. For the GKLaWAC RAP area, only two Traditional Owner fires have been recorded in this spatial layer, both dated 31 March 2021. One was 2.6ha in area, the other 4.4ha. Given the limited data and small spatial extent of the reported Traditional Owner fires in this spatial layer, in this chapter we focus only on patterns of occurrence and extent of wildfires and prescribed burning.

Over the past 91 years, since 1930, much of the GKLaWAC RAP area (25,770km^2) has been burnt at least once by wildfires or prescribed burning, with many areas having been burnt multiple times. Looking at the total area burnt by all fire events, wildfires have burnt more than three-times the amount of land as prescribed burns. Wildfires burnt 17,548km^2 (at an average rate of 251km²/year) of the GKLaWAC RAP area in the 70 years of records for the 20th century (1930–1999), compared to 16,111km² (767km²/year) for the following 21 years (2000–2020) (Figure 5.1). The slightly greater extent of burnt area from the 20th century is largely a result of the huge wildfire event of 1939—the Black Friday fires—that burnt 9031km², just over half the total area burnt in 1930–1999 (Figure 5.2). The average rate at which the landscape has burnt by wildfires in the 21st century is triple that of the 20th. In contrast, the area burnt by prescribed burning in the 20th century was nearly double that of the 21st century, but the latter has done so at nearly twice the average rate of the former (6720km² (96km²/year) versus 3637km² (173km²/year), respectively; Figure 5.1).

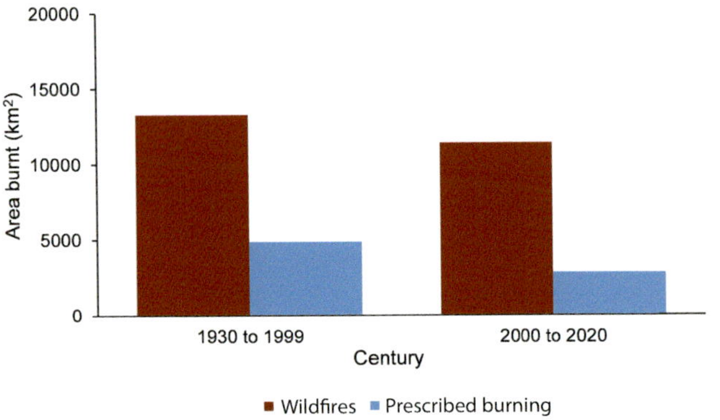

Figure 5.1. Extent of the GKLaWAC RAP area burnt (km²) in the 1900s (1930 to 1999) and 2000s (2000 to 2020). Brown bars represent fires caused by wildfire events, blue bars by prescribed burns. All data sourced from https://discover.data.vic.gov.au/dataset/fire-history-records-of-fires-primarily-on-public-land. The total area burnt per century is cumulative, that is, the area burnt each year within the GKLaWAC RAP area is summed across all years for each century—in this case, some locations might have been burnt by wildfire or prescribed burns more than once.

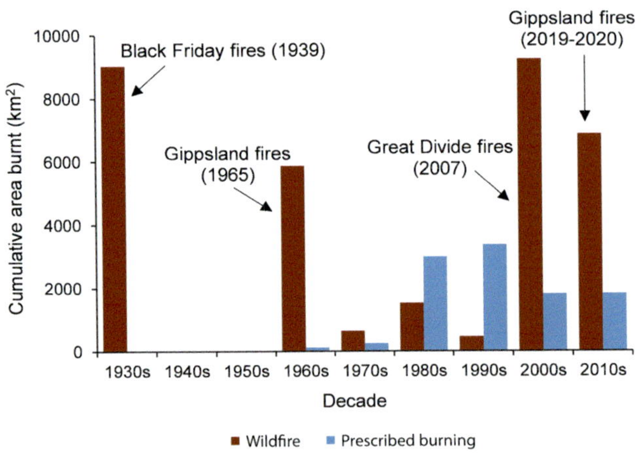

Figure 5.2. Extent of the GKLaWAC RAP area burnt (km²) each decade from 1930 to 2020. The four largest fire events are indicated by arrows to their corresponding decades. The total area burnt per decade is cumulative, that is, the area burnt within the GKLaWAC RAP area each year is summed across all years for each decade—in this case, some locations might have been burnt more than once.

The spatial extent of burnt landscape in the GKLaWAC RAP area has fluctuated over the past nine decades, with the pattern of spikes in wildfire extent being due primarily to individual, large-scale wildfire events (Figure 5.2). The decade with the largest area of land burnt was the 1930s, when the Black Friday fires raged across the landscape (area burnt in 1939 = 9030km², almost 100% of the total area burnt by wildfires in that decade). This was followed by the 2000s and the Great Divide fires (extent burnt in 2007 = 6841km², which is 74% of the total area burnt by wildfires in the 2000s); the 1965 Gippsland fires (extent burnt in 1965 = 4377km², 75% of the total area burnt by wildfires in the 1960s); and the 2019–2020 Gippsland fires (extent burnt = 4992km², 73% of the total area burnt by wildfires in the 2010s).

Wildfires were fewer and smaller in the 1970s, 1980s and 1990s than during all other decades on record (and particularly compared to the past 21 years), occurring in relatively small patches across the GKLaWAC RAP area (Figure 5.3). The 1930s Black Friday fires was the largest wildfire event by area burnt. No wildfires were recorded in the GKLaWAC RAP area in either the 1940s or 1950s, which is why this period is not represented in Figure 5.3. We can only speculate as to why, but this could be due to potential misreporting, or because the wildfires were small, the extents were unknown, or there simply were no wildfires in that 21-year period. Prescribed burns were first recorded for the GKLaWAC RAP area in 1966, and in all 55 years since then. A total of 40% of GKLaWAC land (extent = 10,360km^2) has been prescribed-burned (Figure 5.1). Each year an average of 0.77% of the GKLaWAC RAP area is prescribed burned, with some years seeing up to 2.8% (1987) of land thus burned. The 1980s and 1990s saw more land prescribed burned than any other decade, and also more area burned than by wildfires for these decades (Figures 5.2, 5.3).

In general, both wildfires and prescribed burning have occurred in areas with a lower human footprint (Figures 5.3, 5.4). This means that landscape fires have occurred less frequently in areas with lots of infrastructure and where people live at higher densities—these are often the parts of the landscape that are more easily protected and that contain low amounts of fuel for the fires (i.e., they are often cleared landscapes). These areas with a higher human

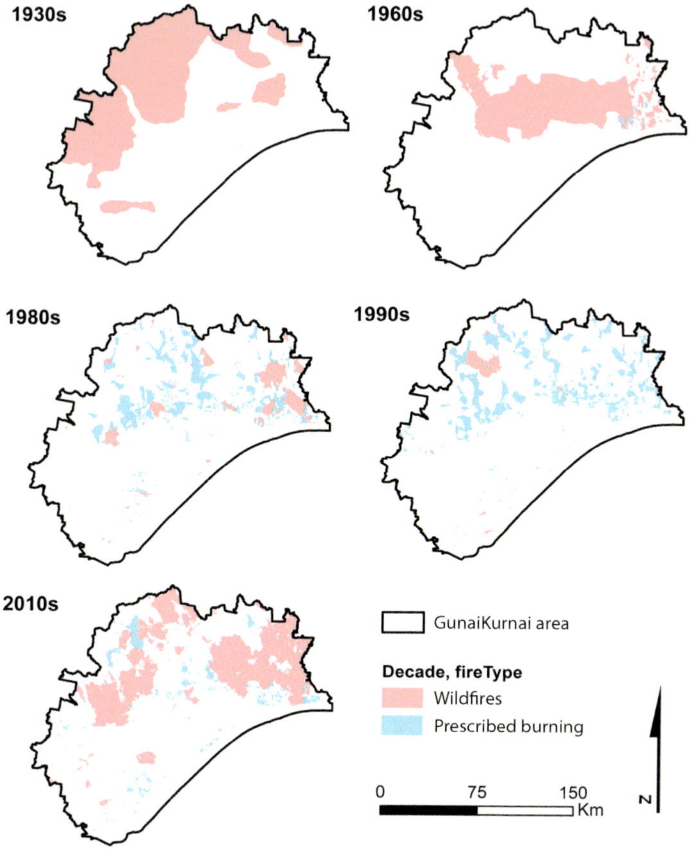

Figure 5.3. Location and extent of landscape fires across the GKLaWAC RAP area by decade, starting with the 1930s. Red shading indicates areas burnt by wildfires, blue areas by prescribed burns. Not shown here is the incidence of multiple burns within each decade.

footprint are also the areas where higher densities of cultural sites have been reported (Figure 5.4, Chapter 6).

Detailed spatial data on the severity of wildfires were not collected until recent times. However, such data do exist for the 2019–2020 Gippsland fires that raged across northeastern GunaiKurnai Country and beyond (Figure 5.5). Five hundred and forty-five square kilometres of the area burnt by this wildfire event was of extreme severity (areas where >20% of the canopy foliage was totally consumed by fire). High-severity burning (where >80% of the canopy foliage was scorched) impacted 980km^2, and low-severity burning (where <20% of the canopy foliage was scorched, much of it being unaffected by the fire) impacted 1066km^2. Conversely, there was a much smaller area of 159km^2 impacted by medium-severity burning (where the canopy is a mosaic of 20–80% unburnt and scorched foliage).

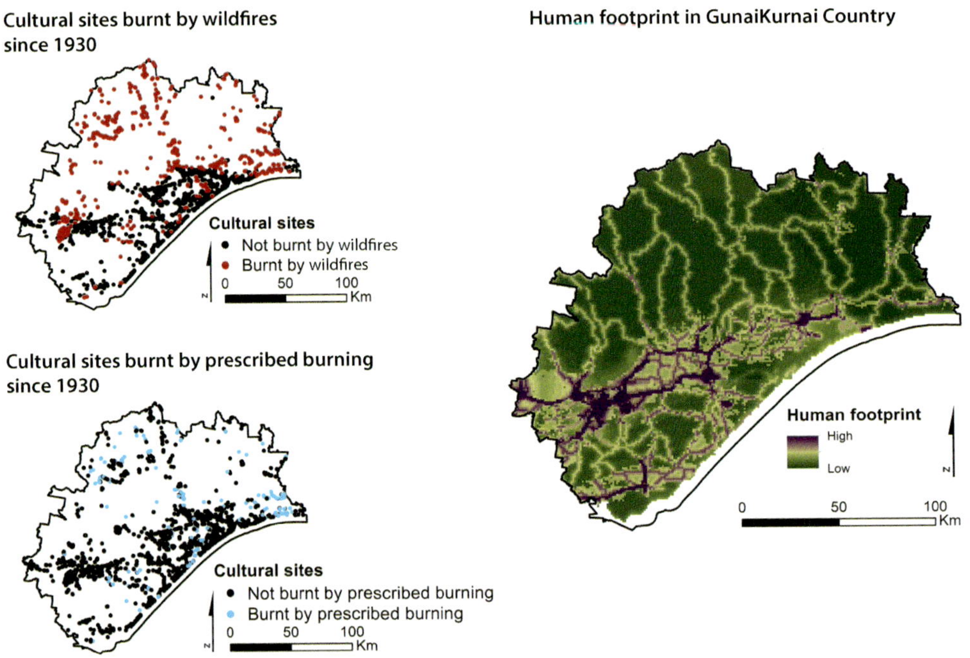

Figure 5.4. Fire history for cultural sites burnt by wildfires (top left) and prescribed burning (bottom left); the right-hand side map shows the relative density of human footprint (roads and other infrastructure, population densities, urban areas) across the GKLaWAC RAP area since 1930. All registered cultural sites are plotted. Data on cumulative human population pressure and footprints were sourced from the Socioeconomic Data and Applications Centre (SEDAC). The spatial (GIS) layer came from the 2018 release and was of a 1km resolution. These data include the sum impacts of the following eight variables: Built-up environments, population density, electric power, infrastructure, crop lands, pastures, roads, railways, and navigable waterways (Human Footprint, 2018 Release: Last of the Wild, v3 | SEDAC Socioeconomic Data and Applications Center | SEDAC (columbia.edu))

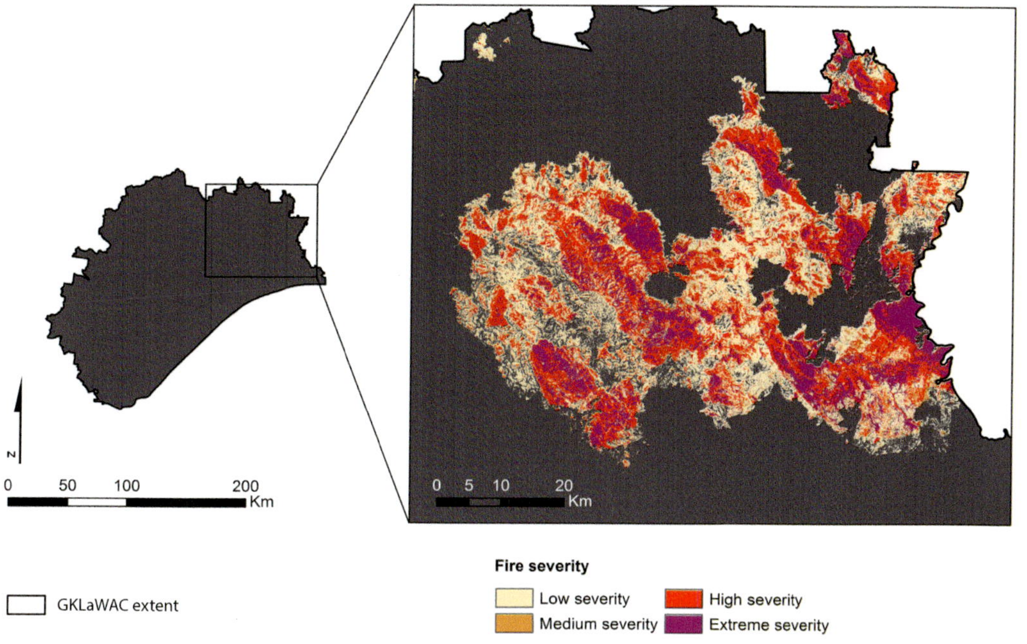

Figure 5.5. The extent of the largest wildfire that burnt across northeastern GunaiKurnai Country in 2019–2020. The spatial data on fire severity was sourced from the Victorian Department of Environment, Land, Water and Planning (https://researchdata.edu.au/severity-map-major-version-10/1459601). The fire severity levels are only shown here for the GKLaWAC RAP area.

CONCLUSION

In summary, small parts of the GKLaWAC RAP area undergo prescribed burning each year. On average, this amounted to 96km^2/year in the 70 years between 1930 and 1999, nearly doubling to 173km^2/year over the past 21 years (to 2020), but the average annual prescribe-burned areas remain low. In contrast, on average 251km^2 of land burned through wildfire each year in the 70 years between 1930 and 1999, and over the past 21 years (to 2020) this increased threefold to 767km^2/year. In other words, two to three times as much land was burnt by wildfires as by prescribed burning each year between 1930 and 1999, but since then this has increased significantly to four to five times. The situation is even more grave when it is remembered that unlike prescribed and cultural fires, wildfires are usually very hot fires of great severity that burn most things in their paths, including to the treetops. Furthermore, wildfires often happen in largely unpredictable, intense bursts during the drier and hotter times of year, the summer months, in contrast to prescribed and cultural burns that can be planned over longer timeframes to minimise risk. The landscapes most affected by wildfires are those of greater distance from high human footprint. Understanding these spatial patterns of wildfire extent and severity enables a better understanding of how different parts of the landscape have been disproportionately burnt, and, with this, to better identify areas of high(er) priority for on-the-ground monitoring, assessment, recovery, and post-fire management. Such issues, including those addressing limitations and caveats of the underlying spatial and survey data, are explored in the following chapters.

PART 2
The Distribution of Cultural Sites in GunaiKurnai Country, and How Fires Affect Cultural Materials

Previous page: Cultural Fire is a pathway to restoring traditional meanings and the reclamation of traditional customs. Photograph by Jessica Shapiro, courtesy of GunaiKurnai Land and Waters Aboriginal Corporation.

Chapter 6

Cultural Sites in GunaiKurnai Country

Jessie Buettel, Russell Mullett, Jessie Birkett-Rees, Bruno David, Jean-Jacques Delannoy, Joanna Fresløv, Stefania Ondei, Robert Skelly and Jerome Mialanes

A key dimension of GunaiKurnai ecology and GKLaWAC's Whole-of-Country Plan is that the whole landscape is cultural (see Chapter 1). What, then, are 'cultural sites' if the whole landscape is cultural? For the purposes of this volume, we refer to individual Aboriginal places as 'cultural sites' if they show archaeological evidence of past Aboriginal activities, or if they are associated with oral traditions, while not forgetting that the entire landscape is an artefact of cultural practice and Country in GunaiKurnai ways and management. We stress also that while we report on cultural sites listed in the Victorian Aboriginal Heritage Register (VAHR), the vast majority of sites in GunaiKurnai Country surely remain unrecorded given the limited surveyed areas. In this chapter we examine the role of heritage legislation in shaping the VAHR and consider the locations of cultural sites in relation to surface geology, current vegetation cover and current land use, and what this may mean for fire risk.

WHAT ARE CULTURAL SITES?

Throughout this volume, the term 'cultural site' equates with the 'Aboriginal Place' of the *Aboriginal Heritage Act* 2006, the primary Victorian legislation that protects Aboriginal heritage sites in the State of Victoria. Section 5 of the Act defines an Aboriginal Place as:

(1) For the purposes of this Act, an Aboriginal place is an area in Victoria or the coastal waters of Victoria that is of cultural heritage significance to Aboriginal people generally or of a particular community or group of Aboriginal people in Victoria.
(2) For the purposes of subsection (1), ***area*** includes any one or more of the following—
    (a) an area of land;
    (b) an expanse of water;
    (c) a natural feature, formation or landscape;
    (d) an archaeological site, feature or deposit;
    (e) the area immediately surrounding any thing referred to in paragraphs (c) and (d), to the extent that it cannot be separated from the thing without diminishing or destroying the cultural heritage significance attached to the thing by Aboriginal people;
    (f) land set aside for the purpose of enabling Aboriginal ancestral remains to be re-interred or otherwise deposited on a permanent basis;
    (g) a building or structure.

When cultural sites are recorded, they are commonly divided into two types, based on the kind of information they are known from:

FIRES IN GUNAIKURNAI COUNTRY (ARCHAEOPRESS 2023): 61–87

Archaeological sites. Archaeological sites have material remains from past cultural activities (e.g., stone artefacts, animal bones or shell from food remains, fireplaces or Ancestral remains). It is these physical remains of the activities of the Old Ancestors that archaeologists usually study to try to understand the past.

Story places. Story places are locations for which community members hold knowledge or memories of past events, or cultural significance. That knowledge is sometimes kept confidential by specific members of the community, such as women, men, or Elders, and may be held by individuals, families or larger cultural groups. Sometimes the knowledge contributing to story places can be broadly known and shared by the wider community.

The *Aboriginal Heritage Act* 2006 refers to such knowledge held in oral traditions as 'intangible heritage', which the Act defines in Section 79B as:

(1) For the purposes of this Act, Aboriginal intangible heritage means any knowledge of or expression of Aboriginal tradition, other than Aboriginal cultural heritage, and includes oral traditions, performing arts, stories, rituals, festivals, social practices, craft, visual arts, and environmental and ecological knowledge, but does not include anything that is widely known to the public.
(2) Aboriginal intangible heritage also includes any intellectual creation or innovation based on or derived from anything referred to in subsection (1).

Therefore, while the traditional knowledge held by GunaiKurnai about cultural places is deemed 'intangible heritage', the places themselves are held as Aboriginal Places under the Act. Some sites can be both archaeological (with artefacts left by the Old Ancestors) and story (where oral traditions are held) places.

There are many different types of cultural sites in GunaiKurnai Country (as indeed there are across all of Australia), and while members of the GunaiKurnai community, and trained archaeologists, have long known this, many outsiders have been unaware of the full spectrum of their existence, including many landscape management authorities. Furthermore, while there are probably tens or hundreds of thousands of GunaiKurnai sites across GKLaWAC Country, many lie on or buried under the ground and can thus be hard to see by those not taught to recognise them, or remain completely hidden from view. For this reason, systematic archaeological surveys, the recording of oral histories, and, more rarely, archaeological excavations are undertaken to systematically identify and record the location and contents of cultural sites. Such cultural sites are referred to in the *Aboriginal Heritage Act* 2006 as 'cultural heritage places'. Some sites can be very small, the location of a single artefact for example, or they can be very large, such as a whole ridge that formed a travel route, or a mountain or a rock that has spiritual significance. When recording and analysing cultural sites, it is therefore important to define exactly what kind of place is being referred to.

A record of known cultural sites in Victoria, including those identified by or with Traditional Owners, or recorded during archaeological research or cultural heritage management (CHM) assessments, is collated in the Victorian Aboriginal Heritage Register (VAHR). The VAHR was established under the *Aboriginal Heritage Act* 2006 (Division 2), in which the State Government of Victoria (Department of Premier and Cabinet) is required to establish and maintain a list of all known Aboriginal places and objects. The current VAHR developed from the records established under the Victorian State Government *Archaeological and Aboriginal Relics Preservation Act* 1972, an organisation that was then called the Victoria Archaeological Survey, and exists with multiple purposes, detailed in Section 144A of the *Aboriginal Heritage Act* 2006. The register exists as a place for Traditional Owners to store information about their cultural heritage (Section 144Aa) and to provide information to assist in the protection and management of Aboriginal cultural and intangible heritage (Section 144Ab). Access to the information about cultural sites held in the VAHR is restricted to people granted access by Registered Aboriginal Parties (RAPs) or by the Aboriginal Heritage Council, and members of the State Government Department responsible for administering the Act (First Peoples–State Relations).

We have devoted attention to the language presented in the Act and the associated VAHR because these documents have significant bearing on the definition of cultural sites for heritage management purposes. This particular definition of cultural sites has disproportionate influence on the information that is now available for use in investigations of the diverse types and locations of cultural sites in GunaiKurnai Country. All legislation necessarily essentialises complex concepts; as others have noted, the legislative regimes that underpin Aboriginal cultural heritage management and protection ('protection' being the primary purpose articulated by the *Aboriginal Heritage Act* 2006) are important mechanisms through which rights and responsibilities may be asserted, but formalising definitions around what constitutes cultural heritage can prove a powerful exclusionary force in future definitions of cultural heritage (Porter 2006; Smith 2000). This issue of standardising but also codifying what constitutes a cultural site is not unique to Victoria or to Australia (Coombe and Baird 2015). Over time, the definitions mandated by a given legislation will influence how, where and what type of information is collected about cultural heritage and will thereby create bias in the known record.

The *Aboriginal Heritage Act* 2006 and the *Aboriginal Heritage Regulations* 2018 (which give effect to the Act) identify parts of the landscape deemed under this legislation to be more likely to contain archaeological sites (Regulations, Part 2, Division 3). These areas of 'cultural heritage sensitivity' are presented in the Aboriginal Cultural Heritage Register Information System (ACHRIS: https://achris.vic.gov.au/#/onlinemap), with these parts of the landscape being given preference in CHM assessments. These areas of legislated cultural heritage sensitivity are applied as part of the process in determining whether an archaeological investigation will take place before an area is developed or otherwise altered and, in areas of Victoria where there is no Registered Aboriginal Party (RAP), the only areas of land that are *mandated* to be investigated in CHM assessments are those of cultural heritage sensitivity under the Act. These include areas known to contain cultural sites, such as anywhere within 50m of a registered cultural heritage place (Regulations, Part 2, Division 3, 25), and geographically defined areas deemed likely to contain cultural sites, such as land within 200m of a named waterway (Regulations, Part 2, Division 3, 26). These areas were designated on the basis of research (Canning 2003; Coutts *et al.* 1978; Gaughwin 1981; Tutchener and Kurpiel 2021), but this research was generalised to the entire landscape of the State of Victoria. Together with definition of specific 'high impact activities' associated with infrastructure that compel a CHM assessment (Regulations, Division 5), the Act and Regulations have created a distributional bias in CHM assessments. Areas legislated as having cultural heritage sensitivity are preferentially investigated, and areas with higher infrastructure are also preferentially investigated, resulting in higher numbers of cultural sites being recorded in these areas. This is a bias that we will return to later in this chapter.

The regulation of methods for recording archaeological sites and artefacts has likewise influenced the ways that CHM professionals formulate the survey, excavation and documentation of archaeological sites. The distributional bias has been noted, but the types of information collected during CHM assessments and the terminology used to register these cultural sites in the VAHR also reflect the requirements of the legislation. Cultural sites are recorded on the VAHR using forms that detail and standardise the information required in order to register a place or object, and therefore govern the administrative record of a registered cultural site. The registration of cultural sites requires the identification of one or more 'component' of that site, which effectively categorises cultural sites into one of eleven available options.

The VAHR site categories ('components') are:

- Aboriginal Ancestral Remains
- Aboriginal Cultural Place
- Aboriginal Historical Place
- Aboriginal Object Collection
- Artefact Scatter
- Earth Feature
- Quarry
- Rock Art
- Scarred Tree
- Shell Midden
- Stone Feature

The definition of each VAHR site type ('component') has changed over time. For instance, prior to 2012, an archaeological site with stone artefacts could be registered as an 'Artefact Scatter' or as an 'Isolated Artefact'. In 2012, the Office of Aboriginal Affairs Victoria introduced a new system for describing and registering areas where fewer than ten stone artefacts were found within an area of $100m^2$ as 'Low Density Artefact Distributions' (LDADs) (Spry 2016). LDADs represent an additional category to the Artefact Scatter listed above. This change in definition is problematic. In relation to understanding the distribution of sites across the landscape, it means that sites recorded before and after 2012 are spread into two or more site types, not because the sites are necessarily different in their characteristics, but simply because they were recorded using different terminology. For the purposes of this volume, LDADs have been distinguished from Artefact Scatters; this choice was necessary given the nature of the information at our disposal, rather than being based on the qualitative characteristics of these sites.

The VAHR information consists of a Heritage Register Name and an eight-digit number for each registered cultural site, geographic coordinates identifying the location of each site, and at least one site category assigned according to the categories defined in the VAHR. This is the baseline information through which a cultural site is registered in the VAHR and displayed in ACHRIS. Additional, qualitative information is provided on component forms producing a 'site card', which is the record produced by the person recording the site for the purposes of registration (see the Victorian Aboriginal Heritage Register Form at https://www.firstpeoplesrelations.vic.gov.au/victorian-aboriginal-heritage-register). The amount of detail provided on existing site cards is not uniform; the main differences appear to relate to the year in which a given site card was produced and the person recording the information. The implication of the varied quality of detail provided in the site cards is that we cannot filter the information equally, across all the records, to improve the current categorisation of archaeological sites.

It is also important to note that despite explicit definitions and instructions, the terms used in the VAHR categories, and the ways in which archaeological sites are grouped or divided, are not always uniformly agreed on by all its users. This is the way the existing information about archaeological sites is presented in the VAHR records, and therefore the official VAHR categories and registrations are used in our research. Furthermore, we understand that the VAHR does not view archaeological sites in the same way as a GunaiKurnai person might.

Aboriginal Ancestral Remains. 'Aboriginal Ancestral Remains' are sites where human burials occur.

Aboriginal Cultural Place. 'Aboriginal Cultural Place' is a broad categorisation that refers to any site that has oral traditions or documented Aboriginal meanings. Sites registered under this category can include archaeological sites with recorded Aboriginal meanings, and story places without material (archaeological) evidence of Aboriginal presence or use. This category is typically associated with a particular location, such as the Talking Dog Site/Legend Rock in GunaiKurnai Country; the significance of the place may be related to traditional cultural knowledge associated with the place as much as to material evidence at the site.

Aboriginal Historical Place. The whole of Victoria is a historical place for Aboriginal people, but for the purposes of the VAHR, an 'Aboriginal Historical Place' is defined as a location that has colonial-settler ('historical') associations with Aboriginal people, such as land reserved for Aboriginal people after European settlement, mission and protectorate stations, places relating to significant known individuals or to Aboriginal self-determination. Sites of colonial conflict, including massacre sites, may also be registered as 'Aboriginal Historical Places'. These 'Aboriginal Historical Places' may contain archaeological remains, such as building foundations, or there may be no material (archaeological) remains, such as at most and possibly all massacre sites. These places may be identified and recorded through oral histories provided by Traditional Owners, by archival documents or archaeological research. This category is wide-ranging and can be used to record places relating to a single event, repeated events, short-term or long-term event. On the registration form for an 'Aboriginal Historical Place', five broad time periods are defined: pre-1830, 1830–1900, 1900–1968, post–1968 or Unknown. These divisions are made on the basis of major social and legislative changes within Australia, including the increased displacement of Traditional Owners after the 1830s; and the Federation of Australia in 1901, in which the colony of Victoria became the State of Victoria within the Commonwealth of Australia. The date of 1968 relates to the 1967 Referendum, in which Section 51 of the Australian Constitution was changed to finally include Aboriginal and Torres Strait Islander people on the Australian census.

The term 'historical' is here used in a problematic way, as history is a continuous process that began prior to the colonial period. There is a tendency in much, but not all, of Australian Indigenous archaeology to differentiate between 'Indigenous' archaeology and 'historical' archaeology, with the latter used to refer to the period since Western intrusion. This terminological differentiation works to make less visible the presence of Aboriginal peoples and Torres Strait Islanders since Western colonialism, as 'historical archaeology' is often identified as the archaeology of Western settler times into the present, while across Australia Indigenous places are thought of mainly through the notion of 'Indigenous places' and 'Indigenous archaeology'. Differentiating 'historical' (the colonial-to-present period) from earlier places gives the wrong message, a problem of terminology.

Aboriginal Object Collection. An 'Aboriginal Object Collection' is defined as any Aboriginal object either held in private collections, in the collections of local museums or historical societies, or reburied on Country. In cases where objects have been returned to Country, the location recorded in the VAHR will represent the location of reburial. In cases where objects remain in a collection, the location of the custodian of the collection (such as a museum or heritage advisor's address) may be listed. This category was previously known as 'Artefact Collection'.

Artefact Scatter. 'Artefact Scatters' are concentrations of stone artefacts seen on or in the ground. They were produced by the activities of the Old Ancestors during camping, tool production and other activities in the course of daily life. For an archaeological site to be recorded as an Artefact Scatter, stone artefacts must be present, but other kinds of artefacts such as food remains (e.g., animal bone, shell), charcoal or ochre may also occur in the same sites. Although Artefact Scatters are numerous across the landscape, no two sites are identical. Such sites can also provide information about interactions between disparate groups across the landscape (e.g., through the presence of imported stone and materials that may have been traded in) and cultural changes over time (e.g., through changes in artefact types or in the technologies employed to make them). An Artefact Scatter can be the result of a single event or activity, or it can indicate places that people returned to over long periods of time. Currently, by VAHR definition an Artefact Scatter consists of a concentration of more than ten stone artefacts within an area $100m^2$ (specifically a $10 \times 10m$ grid), but can be anything from 1m to kilometres long.

Low Density Artefact Distribution (LDAD). As noted above, this type of Artefact Scatter first began to be used in the VAHR in 2012, in order 'to facilitate a streamlined recording process for lower densities of artefacts' (Spry 2016). It therefore only represents more recently registered Artefact Scatters with low densities of stone artefacts (density of ten or fewer artefacts within a $10 \times 10m$ area) that are not in association with other cultural material. Registering an 'Artefact Scatter' requires an extent for an area of artefacts to be recorded, but LDADs do not require that an extent be recorded. Unprovenanced stone artefacts (but not artefacts of any other type) that meet the density criteria outlined above may also be recorded as a LDAD. In the VAHR records, LDADs may be represented by multiple point locations under one VAHR number. More detailed information *may* be available in the VAHR for LDADs relative to Artefact Scatters, given that each artefact must be analysed to register an LDAD, whereas analysis of only a 'representative sample' is required to register an Artefact Scatter.

Earth Feature. 'Earth Feature' is a diverse category including earthen rings, mounds, banks, ditches, canals and trenches. Earthen rings are rare in the VAHR and are generally associated in these records with ceremonial functions. Mounds are distinguished as the result of long-term occupation of a single location and frequently incorporate occupational material such as charcoal, clay heat retainers or stone artefacts. Hearths and ovens are also identified as 'Earth Features', as are 'soil deposits' and 'soil features'. Soil deposits are considered to be accumulations of stratified cultural material, while soil features are defined as the result of soil removal, such as pits or postholes.

Quarry. 'Quarries' are defined as the locations of sources of stone, ochre or mineral pigment(s) that have been procured or extracted by people in the past. Extraction methods can include excavation and breaking, thermal fracturing or flaking, with evidence for the method used typically found on the remaining source stone or material. Quarries exhibit evidence of one or several of the following activities: material extraction, surface collection, transport and reduction or processing. As such, a quarry may also be the location of an Artefact Scatter.

Rock Art. 'Rock Art' sites are defined as places where images have been produced on rock surfaces. These images may be produced either by the addition of pigment(s), sometimes referred to as 'pictograms', or by selective removal of the rock surface, also known as 'petroglyphs'. These images may be isolated or present in multiple locations ('panels') on a rock face. In Victoria, sandstone, granite or limestone are the most common stone types on which Rock Art places were produced. The VAHR draws on the Australian Rock Art Research Association for guidance on recording motifs, forms and designs.

Scarred Tree. 'Scarred Trees' are trees with sections of bark deliberately removed by the Old Ancestors, for the creation of shields, shelters, tools, containers, and bark canoes for transport across lakes and along rivers. The scars on the trees can vary in size, in keeping with the multiple purposes for which the bark was removed, but they are typically regular in shape and often have parallel sides and pointed or rounded ends. Scarred Trees are mature trees most frequently located along rivers and lakes. Some Scarred Trees in Victoria have been carved, containing designs cut into the wood (these are often referred to as 'carved trees', but fall under the Scarred Tree category in the VAHR). Scarred Trees may also refer to trees with cut toe-marks for climbing up the trunk, or cut holes for the extraction of possums or 'sugar bag' (honey) from stingless native bees.

Shell Midden. 'Shell Middens' contain the remains of marine and/or freshwater shellfish harvested by the Old Ancestors. These shell accumulations are called 'middens' in the VAHR records and by most archaeologists. Shell Middens occur in a range of locations: they can be found as layers of shell exposed in dunes and riverbanks, or as shell scatters exposed on surfaces, sometimes through erosion. In addition to shell, which is typically the most common kind of item found in Shell Middens, such sites may also contain fish bone or the bones of marine (e.g., seal) or terrestrial (e.g., kangaroo) animals, usually as food remains. Hearth stones and charcoal from fireplaces may also be found in, or in association with, Shell Middens, as may other kinds of cultural materials.

Stone Feature. 'Stone Features' are defined by their primary construction material. This category takes in several types of features that may have had significantly different functions. 'Stone

Table 6.1. Number of registered cultural sites in the GKLaWAC RAP area.

Site component	Number of sites
Aboriginal Ancestral Remains (Burial)	10
Aboriginal Ancestral Remains (Burial) and Aboriginal Historical Place	1
Aboriginal Ancestral Remains (Burial) and Artefact Scatter and Shell Midden	4
Aboriginal Ancestral Remains (Burial) and Shell Midden	1
Aboriginal Ancestral Remains (Reinterment)	1
Aboriginal Cultural Place	1
Aboriginal Historical Place	13
Aboriginal Historical Place and Artefact Scatter	1
Artefact Scatter	1645
Artefact Scatter and Aboriginal Ancestral Remains (Burial)	6
Artefact Scatter and Earth Oven	1
Artefact Scatter and Grinding Grooves	2
Artefact Scatter and Hearth	4
Artefact Scatter and Hearth and Shell Midden	1
Artefact Scatter and Hearth and Subsurface	1
Artefact Scatter and Quarry	3
Artefact Scatter and Rock Art	1
Artefact Scatter and Rock Art and Subsurface	2
Artefact Scatter and Shell Midden	145
Artefact Scatter and Shell Midden and Subsurface	2
Artefact Scatter and Subsurface	37
Bora Rings	1
Grinding Grooves	20
Grinding Grooves and Quarry	1
Low Density Artefact Distribution	163
Mound	1
Object Collection	34
Quarry	10
Scarred Tree	387
Shell Midden	195
Shell Midden and Subsurface	4
Total sites	**2698**

Features' may represent way- or boundary-markers, such as cairns, or may be related to resource procurement and production, such as fish or eel traps, or grinding grooves used to sharpen stone implements. Subdivisions within this category include the term 'Stone Arrangement', to be used for a feature that was intended for a ceremonial function, and the term 'Stone Structure', used to designate a feature likely to have been a dwelling or shelter. Such dwellings may be those constructed by the Old Ancestors in the more distant past, or stone buildings associated with Aboriginal people during colonial times.

Table 6.2. List of the types of site components found in the GKLaWAC RAP area and the number of sites in which they occur. Some sites may have more than one component.

Site component	Number of sites in which it occurs
Aboriginal Ancestral Remains (Burial)	22
Aboriginal Ancestral Remains (Reinterment)	1
Artefact Scatter	1827
Aboriginal Cultural Place	1
Aboriginal Historical Place	15
Earth Feature	53
Low Density Artefact Distribution	377
Object Collection	96
Quarry	14
Rock Art	3
Scarred Tree	388
Shell Midden	353
Stone Feature	23
Total	**3173**

CULTURAL SITES IN THE GUNAIKURNAI RAP AREA

We now investigate the locations of registered sites within the GunaiKurnai RAP area using the VAHR records. There are 2698 registered cultural sites in the GunaiKurnai RAP area, including from each of the 11 major VAHR site types: Aboriginal Ancestral Remains, Aboriginal Cultural Place, Aboriginal Historical Place, Aboriginal Object Collection, Artefact Scatter (including LDAD), Earth Feature, Quarry, Rock Art, Scarred Tree, Shell Midden, and Stone Feature (Table 6.1). There are many registered cultural sites in the GunaiKurnai RAP area that contain more than one site component type (Table 6.2). By far the most numerous site types are Artefact Scatters, Low Density Artefact Distributions, Scarred Trees, and Shell Middens (Tables 6.1, 6.2).

WHERE ARE THE CULTURAL SITES IN THE GUNAIKURNAI RAP AREA FOUND?

Cultural sites are fragile and often very susceptible to damage by people and environmental disturbance, especially when sites are exposed. Cultural sites are particularly subject to damage or destruction when events operate over large spatial scales and impact the landscape quickly, such as wildfires. It is important to understand and recognise not only where these cultural sites are located, but also the characteristics of the landscapes that surround them, so that we can better understand the impacts that disturbances like fires will have on site and artefact preservation, and their potential destruction. Post-fire management decisions can also lead to the unintentional damage of cultural sites if they are not recognised, as documented by a number of post-fire reports over the past 21 years (e.g., Fresløv 2004; Skelly *et al.* 2019).

The 2698 registered cultural sites are distributed across the entire length and breadth of the GunaiKurnai RAP area (Figures 6.1, 6.2). There are, however, distinct locations with higher densities of registered cultural site types and artefacts, as highlighted by areas of dark red in the density plot on Figure 6.1. These high-density clusters of a number of site types are located along the Gippsland Lakes region (Lakes Entrance), and associated with the towns of Traralgon and Morwell—unsurprisingly, areas of high concentrations of people and urban infrastructure today.

Although the registered cultural sites are distributed across the entirety of the GunaiKurnai RAP area, it is also clear that different site and artefact types have been found in some areas and not in others, with a major pattern being of very few site types in the north of the GunaiKurnai RAP area (Figure 6.1). Artefact Scatters (surface) are the most numerous and widespread of all site types. Scarred Trees are mainly known from closer to the coast, with the occasional occurrence in the north and southwest, while Shell Middens are mainly found near the coast (Figure 6.2). Nevertheless, it is also clear that the known distribution of at least some site types is largely a function of where surveys have and have not taken place.

DISTRIBUTION OF REGISTERED CULTURAL SITES BY SURFACE GEOLOGY (ROCK TYPES AND SURFACE SEDIMENTS)

The GunaiKurnai RAP area covers most of the eastern part of Victoria and a wide range of geologies, ranging from Cambrian to Quaternary deposits. This varied geological landscape offers a large selection of rocks suitable for the manufacture of flaked and ground stone tools and mineral pigment sources. People would have had access to those parent rocks, especially in areas where they are exposed by fault-lines. While the availability of raw materials often reflects the locally outcropping lithologies, other rock types would also have been available from secondary deposits, such as where creeks and rivers have transported eroded rocks along riverbeds across the landscape. Table 6.3 lists the most common rocks present in the GunaiKurnai RAP area by type and age. This list can be used in conjunction with Figure 6.3 to identify their locations in the GunaiKurnai RAP area. More detailed information can be gathered by looking at the rock unit code on the geological maps provided in the Table 6.3 references.

Broadly speaking, there are four key rock types found in the GunaiKurnai RAP area: sedimentary (making up 57% of the study area), igneous (11%), metamorphic (3%) and regolith (21%) (the remaining 8% is made up of water bodies) (Figure 6.3, Table 6.3). Sedimentary rocks are geologically the most common across the GunaiKurnai RAP area, in particular those of the sub-type siliciclastic (silica-rich non-carbonate rocks, which can include sandstone or quartzite, for example). Regolith (loose sediments above the bedrock) is the second-most common, with much of this attributed to the sub-type colluvium (11%) and roughly equal proportions of the other two sub-types (alluvium (6%) and coastal dunes (5%)). Even though the area covered by regolith (5474km^2) is substantially less than that covered by sedimentary rock (14,586km^2) as delineated on the geological map, it is where the majority (60%) of the registered cultural sites are located (Figure 6.3).

Of special relevance to discussions of heritage and the impacts of wildfires is a specific consideration of rock shelters. Caves and rock shelters are found in many parts of GunaiKurnai

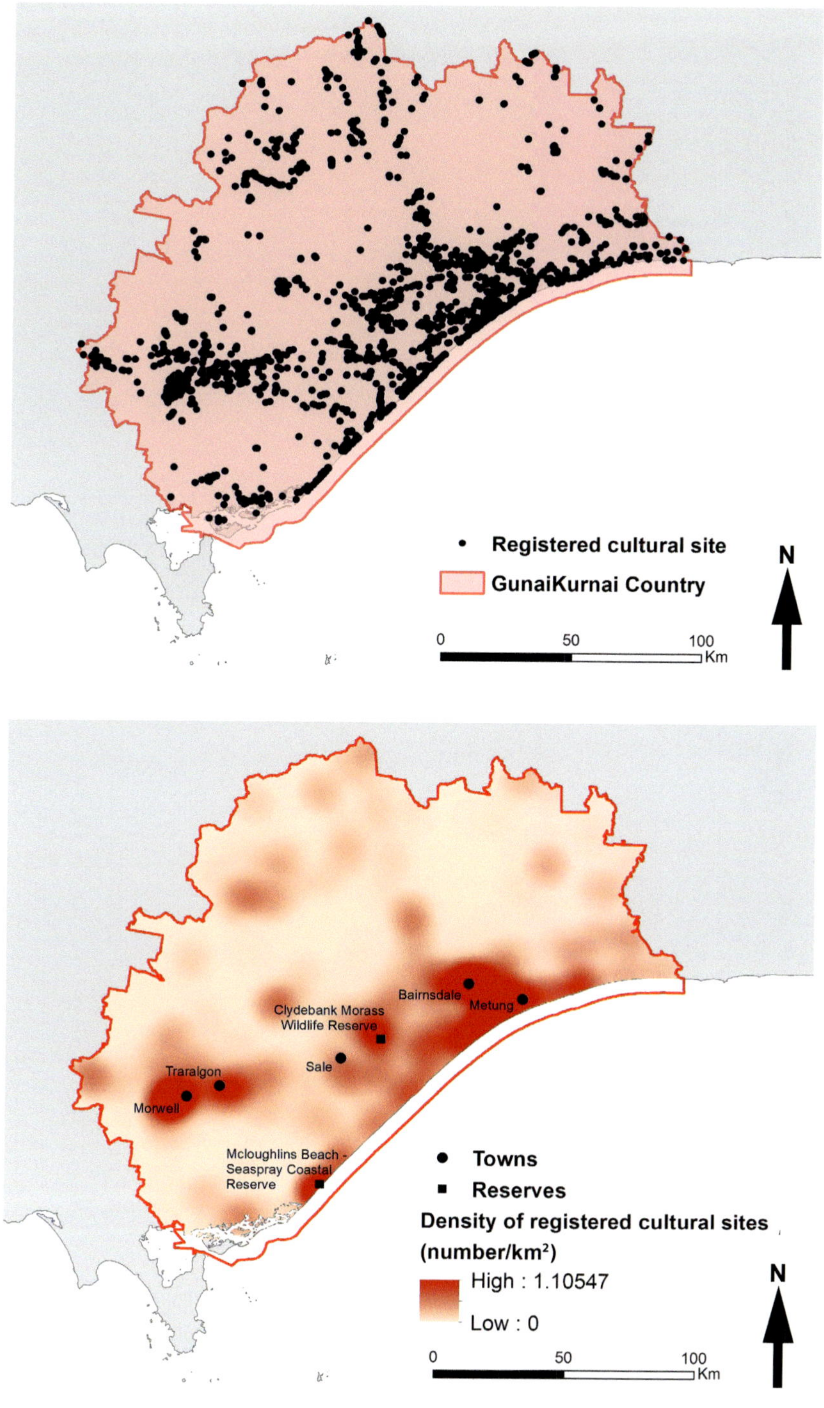

Figure 6.1. Location of all 2698 VAHR-registered cultural sites in the GKLaWAC RAP area. Top: Individual registered cultural site locations plotted. Bottom: Density map expressed as number of registered cultural sites per km^2.

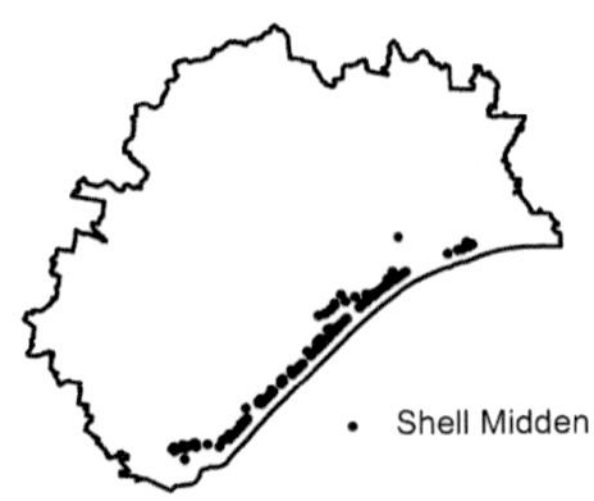

Figure 6.2. Location of different site component types across the GKLaWAC RAP area. Each dot represents a registered cultural site that contains that component type. Some registered cultural sites will have more than one component type (the counts of each component type can be found in Table 6.2).

Table 6.3. Major lithologies of the GKLaWAC RAP area.

Lithology	Age (Unit code)	Geological map references
Igneous rocks		
Granite	Upper Devonian (Dug)	VandenBerg 1997a, 1997b
	Middle Devonian (Dmg)	VandenBerg 1997a
	Lower Devonian (Dlg)	VandenBerg 1997a
	Silurian-Devonian (Sdg)	VandenBerg 1997a
Basalt	Tertiary Basalt (Tvo)	VandenBerg 1997a, 1997b
Metamorphic rocks		
Quartzite	Lower Devonian (Diqw)	VandenBerg 1997a
	Ordovician-Silurian (O-S)	VandenBerg 1997a, 1997b
Gneiss	Middle and Lower Ordovician (Osn)	VandenBerg 1997a
Hornfels	Lower Devonian (Disy)	VandenBerg 1997a
Sedimentary rocks		
Chert	Lower Devonian (Disk)	VandenBerg 1997a
	Lower Silurian (Sle)	VandenBerg 1997a
	Upper Ordovician (Ouu)	VandenBerg 1997a
Limestone	Lower Devonian (Sjp)	VandenBerg 1997c
	Lower Devonian (Dir)	VandenBerg 1997a
	Lower Devonian (Dia)	VandenBerg 1997a
Mudstone	Upper Devonian (Das)	VandenBerg 1997a, 1997b
	Lower Devonian (Diwn)	VandenBerg 1997c
	Lower Devonian (Sj)	VandenBerg 1997c
	Lower Silurian (Sc)	VandenBerg 1997a, 1997b
Sandstone	Lower Cretaceous (Kls)	VandenBerg 1997c
	Upper Devonian (Das)	VandenBerg 1997a, 1997b
	Lower Devonian (Diwn)	VandenBerg 1997c
	Lower Devonian (Sjp)	VandenBerg 1997c
	Lower Devonian (Sj)	VandenBerg 1997c
	Lower Devonian (Dll)	VandenBerg 1997c
	Lower Silurian (Sjn)	VandenBerg 1997c
	Lower Silurian (Sc)	VandenBerg 1997a, 1997b
Siltstone	Cretaceous (Kls)	VandenBerg 1997c
	Lower Devonian (Dll)	VandenBerg 1997c
	Lower Devonian (Sjp)	VandenBerg 1997c
	Lower Silurian (Sjn)	VandenBerg 1997c
Silcrete (silicification of regolith)	Extremely common across the GKLaWAC RAP area	

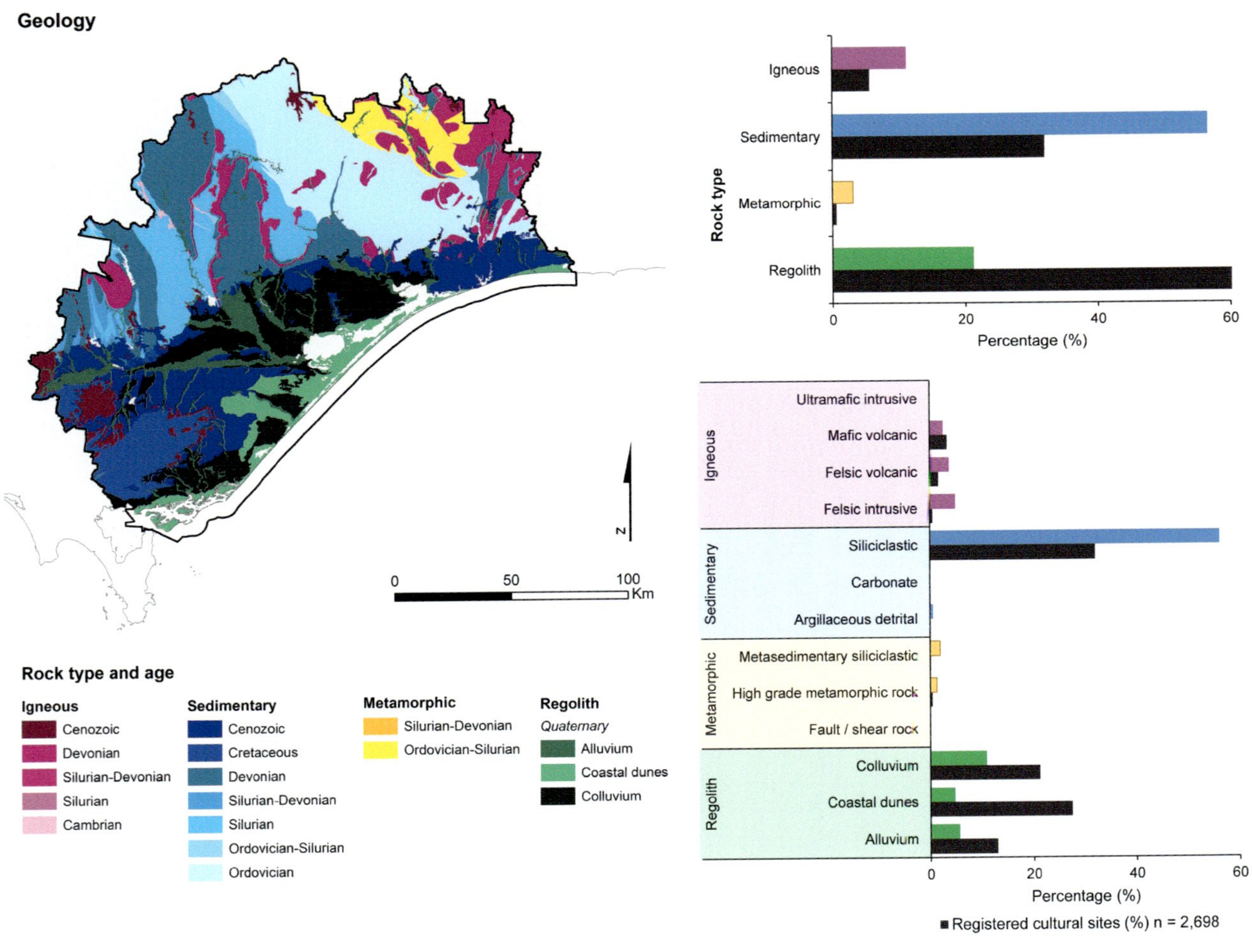

Figure 6.3. Map showing the surface geology of the GKLaWAC RAP area. The top-right graph shows the proportion of rock type for the GKLaWAC RAP area (colour-coded based on the left map and legend) and the proportions of registered cultural sites (dark grey bars) by rock type. The bottom-right graph shows a more detailed breakdown of the distribution of registered cultural types by lithology. Data sourced from: Geoscience Australia (2012) Surface Geology of Australia, 1:1,000,000 scale, 2012 edition. Bioregional Assessment Source Dataset (http://data.bioregionalassessments.gov.au/dataset/8284767e-b5b1-4d8b-b8e6-b334fa972611).

Country, and in a range of lithologies including limestone, conglomerate, and granite. The vulnerability of rock outcrops to the impact of heat across the landscape, and of the rock surfaces and the marks left on them by the Old Ancestors—such as rock art and stone tool quarrying scars—depends on a number of factors. Rock outcrops such as rock pavements, cliffs and boulders are at once topographic features of the landscape, such as open sites, rock shelters and caves, and lithological formations with particular mineralogical, chemical and structural characteristics. Understanding both a rock outcrop's position in the landscape and its geological make-up is fundamental to understanding its vulnerability to fires and, with this, to identifying which conservation and management options are the most appropriate.

Two types of heat generation causing thermal stress on the rock can be differentiated: 1) day–night temperature variations which cause 'thermoclastic exfoliation' (physical weathering through repeated very low-temperature heating and cooling); and 2) exposure to high heat from fires, which causes rocks to fracture and shatter. Thermoclastic exfoliation is the superficial separation ('desquamation') of the outer tier of the rock through very low heat. All the rock types (granite, limestone, quartzite, and so forth) in GunaiKurnai Country are subject to this type of effect. Thermoclastic exfoliation is caused by the shrink and swell of moisture and salts seeping through the rock during day–night temperature changes. This causes sub-centimetre-deep, scale-like surface layers of rock to expand and contract, resulting in the surface of the rock breaking off or 'exfoliating'. Exfoliation spalls are small, centimetre(s)-long thin lenses of rock that retain on their dorsal (outer) surfaces the features of the original rock surface. Thermoclastic exfoliation of rock walls from the heat of the sun and cool of the night only affects the outer surface of the rock. The impact of solar heating depends on the degree of temperature oscillation as well as on the location of the rock. It is strongest in deserts and high mountain areas where day–night temperature variations tend to be at their greatest. In the Southern Hemisphere, rock escarpments facing north, and in the Northern Hemisphere facing south, are often directly subjected to greater levels of solar radiation, and therefore their surfaces can be more exfoliated. For this to happen, the temperature change must be greater than 2 °C per minute, being the critical speed of temperature change needed for the elastic response threshold of the rock to be exceeded and for micro-cracks to appear (Bahr *et al.* 1986; Germinario *et al.* 2015). Through repeated day-night expansion and contraction of the rock surface, especially along fine cracks and, in summer, when the maximum heat of the sun hits the rock, the rock surface becomes fragile until the small and shallow spalls break and then fall off the rock-face through gravity. This separation of the exfoliation spalls from the rock face is a form of fatigue of the rock's crystalline structure. The micro-fissures that form during thermoclastic exfoliation become multiple points of weakness enhanced by weathering, the penetration of humidity, and gravitational force. Although the temperature variations occur on a daily basis, their repeated effects are felt over extended periods of time, and exfoliation spalls take a while to fully separate and fall off, given that they are small and the pull of gravity is therefore relatively minor. It is of interest to note that this phenomenon can preferentially affect the darker colours of the rock, which absorb more heat and therefore reach the elastic response threshold of the rock more quickly. Dark surface colours may be inherent to a rock, or they may be caused by environmental transformations such as the growth of lichen-algal biofilms, manganese flows, and the like, or by the actions of people such as through the deposition of soot from campfires (Figure 6.4). Everything else being equal, thermoclastic exfoliation will tend to affect dark rock surfaces more than light-coloured surfaces.

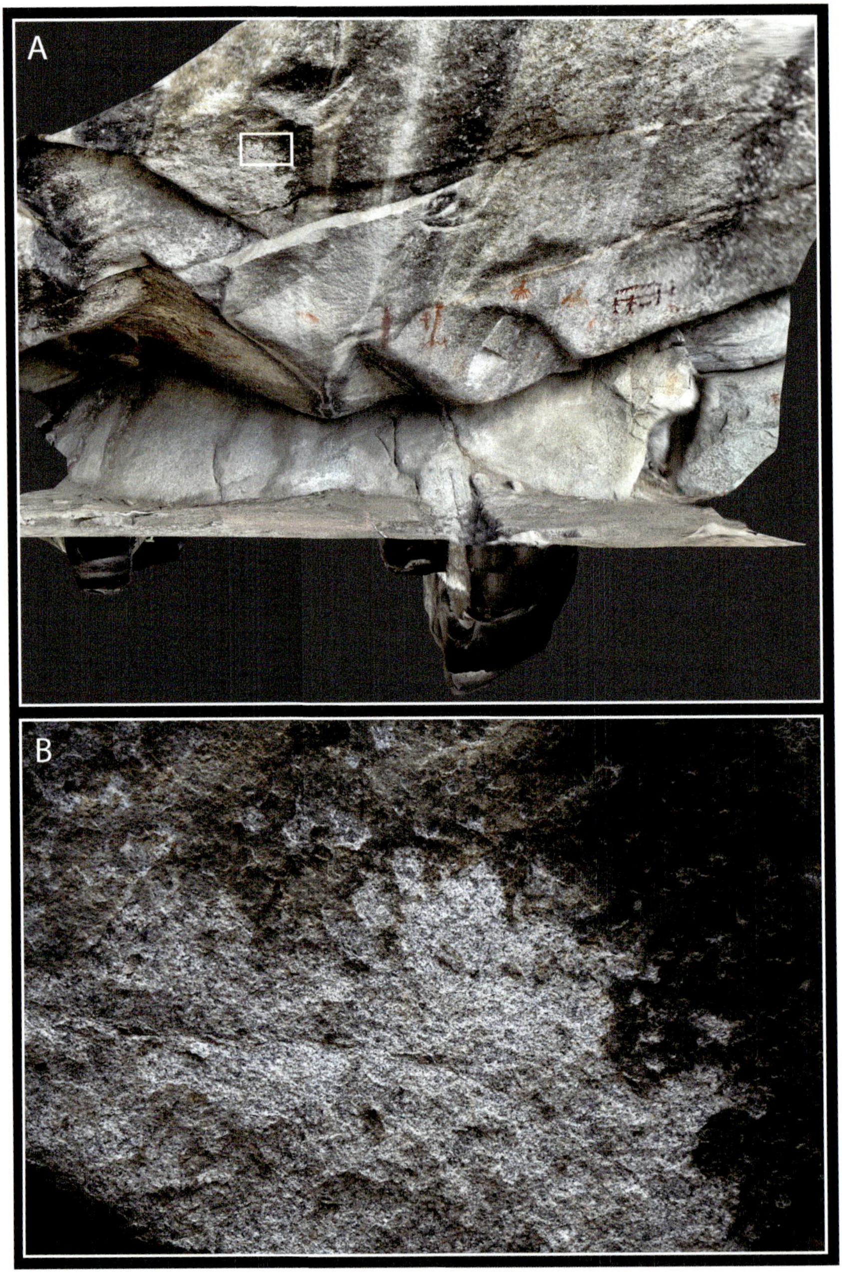

Figure 6.4. Thermoclastic exfoliation on a granite wall blackened by fires at the Chuchuwayha rock shelter (British Columbia, Canada). A: On the walls of the overhang of the rock shelter, there are streaks of soot from past fires. The darkened walls have made them more susceptible to thermoclastic exfoliation. The rock wall faces south and is directly exposed to solar radiation. The site is located in mountainous terrain with low levels of humidity, and high day-night temperature contrasts. B: Thermoclastic exfoliation of the darkened rock wall (photo and 3D model by Jean-Jacques Delannoy, montage by Jean-Jacques Delannoy and Bruno David).

The effects of fires on archaeological rock shelters and cave walls are different to those of thermoclastic exfoliation. Heating through fire occurs more rapidly and causes greater transformations both on the morphology of the rock surface and in the physico-chemical properties of the rock. The impacts of fires tend to be localized, remaining close to the source(s) of heat. In this respect, the uncontrolled, high heat and horizontally as well as vertically (incorporating the vegetation canopy) widespread nature of wildfires causes large sections or even entire rock outcrops to be affected. Fires impact most rocks in open air sites, cliffs, rock shelters and even caves.

A rock's resistance to fire depends on its shape, composition, degree of homogeneity of physical structures across all directions ('isotropy'), and degree of cracking. Paradoxically, rock that is able to differentially move in different directions ('anisotropy'), that already has fissures, that is layered, and/or that is capable of changing shape ('plastic') is less sensitive to strong rises in heat. Compact and hard rocks such as magmatic rocks like granite and lava (which formed from molten substances beneath the Earth) and metamorphic rocks such as gneiss, quartzite and marble (which geologically formed through high-intensity heat and/or pressure) will be more vulnerable to the heat of landscape fires. Sandstone and carbonate rocks also behave in their own ways in the face of fire. When forests burn, the high temperatures reached (typically 800 to 1200 °C), especially at the start of a fire when high levels of oxygenation take place, will suddenly overheat the surface of the rock, which will expand faster than its interior. This produces very strong tensions between the heated external and unheated internal parts of the rock, causing the rock to shatter. This will be marked by the detachment of 'scallops' of rock that are more or less parallel to the rock surface. When the fire is positioned against the rock wall, such detachments are often curved in shape ('scalloped', as in the shape of a scallop shell). Depending on the ability of the rock to conduct heat, the detached pieces will range from a few centimetres to a few tens of centimetres in maximum thickness. The degree to which a rock breaks in this way is dependent on the intensity of the fire. Thermal shock can be accentuated by brutal cooling, such as when a heated wall is rapidly cooled with water. This can take place when a rock heated by wildfire is suddenly cooled by heavy rainfall. Intense heating of rock shelters can cause sections of the wall to fracture, resulting in overhang collapse and the localised retreat of rock walls. Rocks shattered from thermal shock typically have sharp edges.

In limestone rocks, a triple effect can take place, again depending on the heat intensity: 1) the rock wall can redden in colour (ranging from pink to grey) at temperatures between 250 and 400 °C; 2) The wall can break apart; and 3) once temperatures reach 900 °C, decarbonisation takes place and the rock changes from limestone to lime as carbon dioxide (CO_2) is sequestered from the calcium carbonate ($CaCO_3$) to leave behind the transformed, calcium oxide (CaO) limestone rock (Hanein *et al.* 2021) (Figure 6.5). This transformation is analogous to that described for shell in Chapter 11. During this decarbonation of the original limestone, the remnant calcium oxide limestone rock becomes more powdery than before. This transformation takes place over several centimetres in depth across its surface.

Some heated rocks also often redden when exposed to fire; this is so of quartzites as well as iron-rich limestones (David *et al.* 2017: 417–418). Sandstones, due to their higher porosity, are relatively insensitive to low-temperature fires, but above 600 °C micro-cracks begin to appear. However, the micro-fissures can expand as water enters the cracks, salts accumulate and weathering progresses.

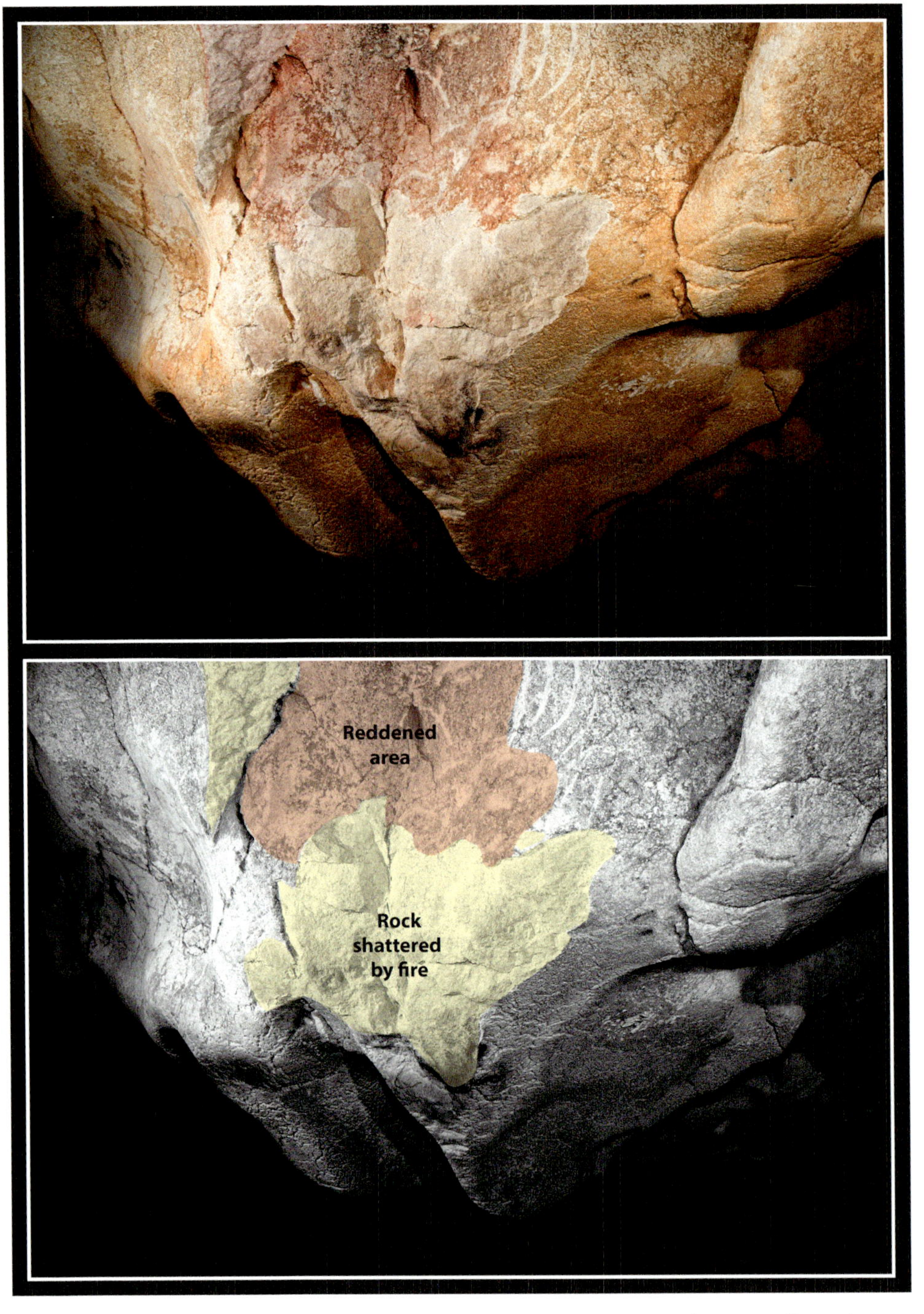

Figure 6.5. Limestone wall reddened and scalloped by adjacent fires, Megaloceros Gallery, Chauvet Cave (France). Several areas in the Megaloceros Gallery have walls reddened and scalloped by Upper Paleolithic fires constructed by people. These fires post-date cave bear scratch marks on the walls, but pre-date the torch marks visible in the lower part of the scalloped wall (photo by Julien Monney, montage by Jean-Jacques Delannoy and Bruno David).

Above ground, the greatest impacts of heat on rocks are those caused by forest fires and, usually to a lesser degree, by fireplaces positioned along cliff-faces, rock shelter walls and near ceilings, and in rock cavities. Forest fire and campfire heat are each capable of reaching temperatures of 800 to 1500 °C, although in Australian conditions campfires usually remain below 900 °C. Therefore, both are capable of shattering nearby rock (Figure 6.6). Herbaceous, small brush and grass fires have less impact on rock surfaces due to their lower temperatures. They also tend to be of significantly shorter durations than forest fires, which both reach tree canopies and burn tree trunks. Campfires near rock walls also often cause the rock to break and alter its matrix and colour.

Both landscape fires and campfires can impact rock shelter overhangs and cave ceilings through heat and smoke (Figure 6.7). For example, the *c.* 21,000 to 36,000 year-old archaeological site of Chauvet Cave in France has rock surfaces that appear to have been scalloped and reddened by the heat of fires deep in the cave. To determine how this had happened, detailed analyses were made of the affected rock at Chauvet Cave, and experiments were carried out in a commercial rock quarry nearby. The aim of this research was to find out the fire intensity required to redden and scallop the limestone rock (Mindeguia *et al.* 2013; Salmon 2019).

First, thermoluminescence analysis was undertaken on the affected Chauvet Cave wall, revealing that temperatures of 250–400 °C had been reached on the walls, causing them to redden (Brodard *et al.* 2014). These conclusions were then tested by undertaking experiments on a limestone wall in a commercial quarry a short distance from the site. During these experiments, the rock began to break off at 600 °C. Elsewhere, experiments also showed that in quartzite rock, once a fire reached a temperature of 800 °C the walls began to break apart after 10–15 minutes of heating (Ancel and Py 2008).

Beyond the mechanical processes that rapidly shape or re-shape (modify) a rock as a fire burns and cools, there are also longer-term effects to consider. These include the physical and chemical modification of the rock, the appearance of micro-cracks, and changes in the colour of rock walls. These longer-term effects especially apply to rock outcrops and rock shelters exposed to the elements. Sometimes, a rock shelter's fire-affected rock surfaces can increase the potential for those surfaces to be further damaged by thermoclastic exfoliation, such as where a fire has caused the rock surface to develop a red or dark patina. The darkened/reddened surfaces cover lighter parent rock, absorbing more heat than they reflect. Exfoliated spalls of rock can detach at the junction between the outer surface heated by solar radiation, and the inner, non-heated rock within. When this happens, the exfoliation caused by solar radiation can also be seen on the parent rock in the colour contrast between the underlying, unaltered and undarkened/unreddened rock exposed by the exfoliation scars and the thin, reddened rock skin on the remaining surface around the scars (e.g. David *et al.* 2017: 417–418).

Detailed study of rock walls and of the shape of open-air rock surfaces would enhance one's ability to determine the conservation measures required of archaeological sites subject to fire in GunaiKurnai Country and beyond. For this to happen, on-site studies and monitoring would be required to properly assess and quantify the state of the rock before and after fires, and for a range of different rock types, topographies and ecosystems (e.g., forests, grasslands, etc.). Such analytical work and monitoring would reveal concrete information by which to better assess the vulnerability of rock outcrops, rock shelters and caves to wildfires as well as

Figure 6.6. Granite boulder decorated with paintings near the town of Osoyoos (British Columbia, Canada), degraded by wildfire in the summer of 2021. The forest fires occurred during a period of prolonged drought and strong heatwaves (+40 °C). The boulder was previously surrounded by highly flammable shrubby vegetation that is particularly sensitive to fire. The fire-scars are up to about 1m in height, corresponding to the height of the burnt vegetation. It is in the upper canopy of the vegetation that the highest temperatures reached (photo by Jean-Jacques Delannoy, montage by Jean-Jacques Delannoy and Bruno David).

Figure 6.7. Limestone wall and ceiling impacted by Upper Palaeolithic campfires, Chauvet Cave (France). The entrance chamber of Chauvet Cave, now sealed by rockfall from the collapse of the cliff above the entrance, retains traces of Upper Palaeolithic campfires. These campfires reddened (pink), darkened (grey) and scalloped the adjacent cave wall. The rock reached temperatures of *c.* 250 °C to turn pink, and 350 °C to turn grey. The scalloping of the rock required higher temperatures. The fire-scalloped scar is here more than 2m high, indicating the intensity of the fire(s) (photo by Jean-Jacques Delannoy, montage by Jean-Jacques Delannoy and Bruno David).

prescribed fires in GunaiKurnai Country, and thereby to further inform management plans across the region's varied landscapes.

Rock shelters that have been damaged or destroyed by wildfires are rarely reported, despite being numerous across the landscape. Unless they had been listed in site registers (such as VAHR) prior to the impacts of wildfires—and this would usually only be the case where surface cultural deposits, flaking of the rock wall such as for the manufacture of stone artefacts, or rock art were known—the effects of wildfires on rock shelters have not been documented for any part of GunaiKurnai Country. Observations on the impacts of landscape fires on rock art and rock art sites are discussed in Chapter 8.

DISTRIBUTION OF REGISTERED CULTURAL SITES BY VEGETATION TYPE

The diverse native vegetation of Australia has been grouped by scientists and representatives of the Australian State and Federal Governments, for management purposes, into 33 major vegetation types and 85 subgroups, as described by the National Vegetation Information System (NVIS: https://www.environment.gov.au/land/native-vegetation/national-vegetation-information-system). For visualisation and simplicity in this chapter, we have grouped these into eight categories based on the similarities between the dominant tree species and type of vegetation. Note that we used the 85 vegetation subgroups for our groupings, as this provides the ability to separate the major vegetation type, 'Eucalypt Forests', into wet versus dry sclerophyll forests. This is an important distinction when exploring fire impacts and prevalence. We created the following eight categories:

1) Rainforest, which can include vine thickets, often has no dominant tree species, and plant communities consist of dominant shrubs from one or more Antarctican (cool temperate) genera such as *Nothofagus*, *Podocarpus* and *Athrotaxis*.
2) Eucalypt Forest—Dry Sclerophyll (tall and open eucalypt forests often with a grassy or shrubby understory).
3) Eucalypt Forest—Wet Sclerophyll (tall, open, and often with a dense broad-leaved and/or fern understory).
4) Eucalypt Woodland (defined by areas where tree crowns are not touching and trees are sparse).
5) Shrubland (dense thickets of vegetation with thick crown cover of low-growing shrubby forms of *Eucalyptus* and *Acacia* and/or dominated by shrub species such as *Melaleuca* and *Leptospermum*. Shrubland includes heathlands, saltbushes and marshes and areas dominated by low, fine-leaved shrub species).
6) Grassland (includes grasses that form distinct tussocks or hummocks, both perennial and annual, as well as some herblands, sedgelands and rushlands).
7) Other (other forests and woodlands, including those dominated by one of either *Acacia*, *Casuarina*, *Callitris*, *Melaleuca*, unclassified native vegetation, naturally bare, regrowth and other native vegetation and 'unknown').
8) Cleared (cleared vegetation, non-native vegetation, and buildings).

For this chapter, we have omitted all waterbodies, lakes, and rivers from our analysis, maps and plots, and used the vegetation spatial layer provided by the National Vegetation Information System (NVIS) for all mapping (https://www.awe.gov.au/agriculture-land/land/

native-vegetation/national-vegetation-information-system/data-products#mvg60). Note that when comparing the numbers of registered cultural sites within each vegetation type, 108 of these fell under the category 'water' on the GIS maps. Where reasonable, these sites (e.g., sites just outside—always <100m from—the coastline) were allocated the closest vegetation type. These sites appear to be offset from their true positions by less than 100m, probably reflecting inaccuracies in their original recording. After doing this, there remained three registered cultural sites located in Lake Glenmaggie that were removed from the analysis, as they remained directly in the water (see below, where we include water as a land-use category).

By far the most common vegetation types in the GunaiKurnai RAP area are the Eucalypt Forests (covering 50% of the total GKLaWAC RAP area). Of the Eucalypt Forests, 35% (9148km^2) are Wet Sclerophyll, and 15% (3777km^2) are Dry Sclerophyll. In contrast, the least widespread vegetation types are Rainforest and Grassland (each represents <1% of the GKLaWAC RAP area). Rainforest areas are scattered throughout the Eucalypt Forests (particularly in the southwest of the study area), and Grassland among the cleared lands closer to the coast in the south (Figure 6.8). Towards the north, contiguous areas of Eucalypt Woodland (10% of the GKLaWAC RAP area) are interspersed among patches of Eucalypt Forest, while to the south along the coast near Lake Wellington, Shrubland predominates (Figure 6.8). Notably, much (just under 30%, 6917km^2) of the GunaiKurnai RAP area towards the southwest corner of the study region has been cleared for infrastructure such as roads, urban areas, and related developments (Figure 6.8).

Even though the Eucalypt Forests are the most common vegetation type, this is not where most of the registered cultural sites are found. Rather, almost half (43%, n = 1157) of all the registered cultural sites in the GKLaWAC RAP area are found in the Cleared vegetation type; that is, in areas that once held native vegetation that have since been cleared for (e.g., infrastructure, agriculture or plantations; Figure 6.8, 6.9). Eucalypt Forests and Eucalypt Woodland contain equally the second-highest number of registered cultural sites (18–20%), despite the smaller area covered by Eucalypt Woodland (2658km^2) compared to the much more widespread Eucalypt Forests (Wet and Dry Sclerophyll = 12,925km^2). Similarly, Wet Sclerophyll and Dry Sclerophyll Eucalypt Forests contain comparable numbers of registered cultural sites (10%), despite Eucalypt Forest—Wet Sclerophyll covering almost triple the area of the Eucalypt Forest—Dry Sclerophyll (9148km^2 vs 3777km^2) (Figure 6.8). Shrubland contains 14% of the total number of registered cultural sites, which is high given its small area (715km^2, or 3% of the total GKLaWAC RAP area), with Rainforest having very few registered sites (n = 2, 0.07%, in 0.27% of the GKLaWAC RAP area).

DISTRIBUTION OF REGISTERED CULTURAL SITES BY LAND USE TYPE

There are five primary types of land-use in Australia as defined by The Australian Land Use and Management Classification (ALUM; https://www.awe.gov.au/abares/aclump/land-use/alum-classification). Each type is distinguished in order of increasing levels of human development, management, intervention, and their impacts on the environment. The classes and examples of each are:

1) Conservation and natural environments (including nature reserves, wilderness areas, national parks, protected land and other minimal-use areas such as rehabilitation and residual native vegetation cover).

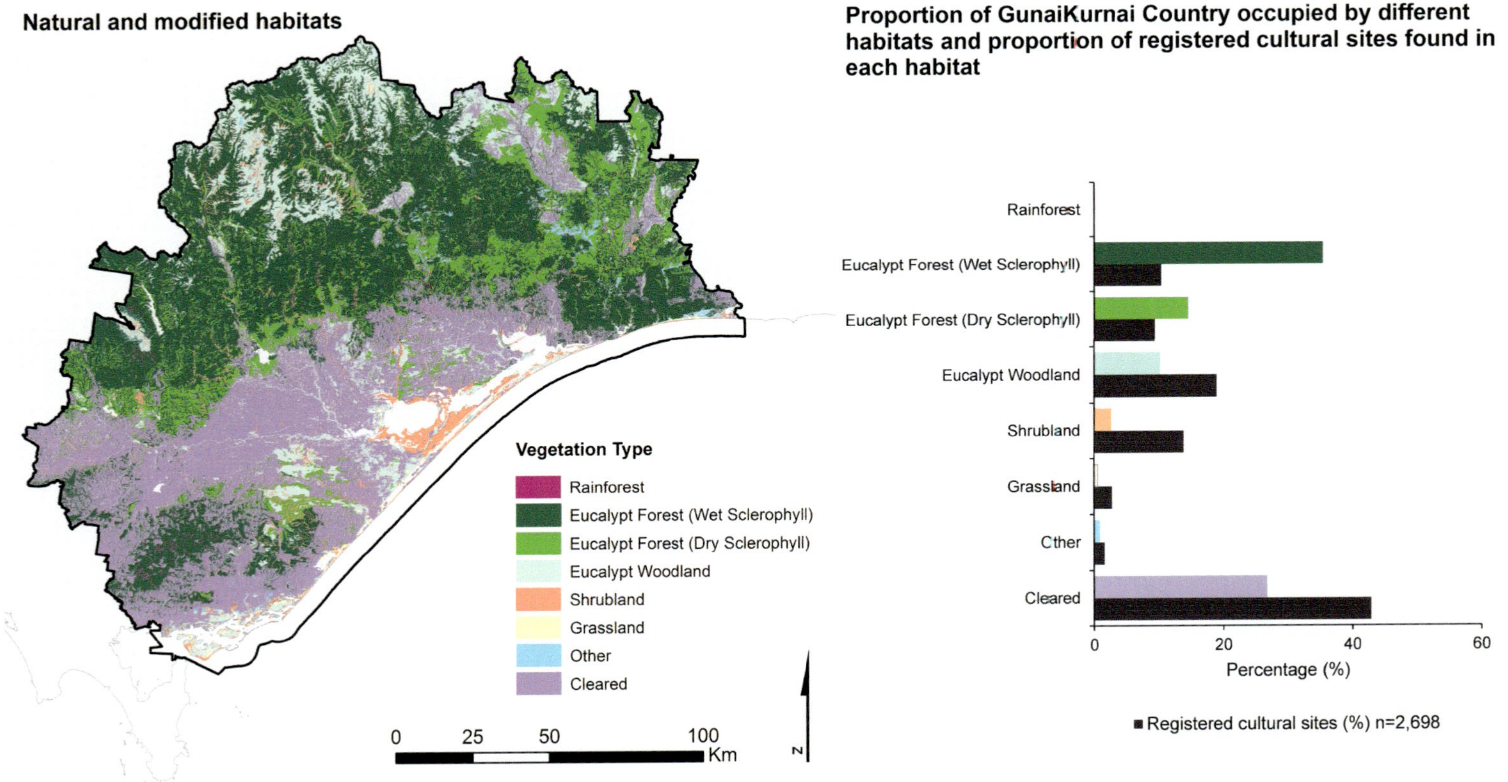

Figure 6.8. Left: Map showing the distribution of each of the seven broad vegetation types in the GKLaWAC RAP area. Right: Proportion of land covered by each vegetation type (colour-coded based on the left map and legend), and of registered cultural sites (dark grey bars), in the GKLaWAC RAP area. These data on vegetation types were sourced from the National Vegetation Information system (NVIS) database (https://www.environment.gov.au/land/native-vegetation/national-vegetation-information-system). The most recent year for these data was used (version 6, 2020), at a resolution of 100 × 100m.

2) Production from relatively 'natural' environments (such as grazing native vegetation, and native forests used for the production of wood and other products).
3) Production from agriculture and plantations (including production from both dryland and irrigated systems; grazing modified pastures; hardwood and softwood plantations; horticulture; cropping and land in transition).
4) Intensive uses (intensive horticulture and agriculture, manufacturing and industrial, residential/urban, transport infrastructure, utilities, and mining).
5) Water/Other (water bodies, rivers, lakes, marsh or wetlands, estuary, or coastal waters).

We have omitted all waterbodies, lakes, and rivers from our analysis, maps and plots in this chapter and used the land-use spatial layer provided by the Australian Bureau of Agricultural and Resource Economics and Sciences (ABARES).

Using the four categories of land-use listed by The Australian Land Use and Management Classification (1–4 above), 75% of the GKLaWAC RAP area falls under some type of Production land or Intensive use (Categories 2–4), compared with only 20% that is under Conservation tenure or deemed to be 'natural' environment (Category 1; Figure 6.9). The remaining 5% of the GKLaWAC RAP area is classified as 'other' (including water bodies and rivers; see Category 5 above). Of the land that falls under Production land or Intensive use (Categories 2–4), Production from relatively 'natural' environments makes up the most of the GKLaWAC RAP area (37%), closely followed by Production from agriculture and plantations (33%) and Intensive uses (the highest category of human impact) (6%). In contrast, 66% of all registered cultural sites are found in the Production and Intensive use areas that covers 75% (19,322km^2) of the total land area, with the other 34% found in the Conservation and 'natural' environments that cover 20% (5075km^2), a much smaller area of the GKLaWAC RAP area. Intensive use areas (6% of the GKLaWAC RAP area) contain a high density of registered cultural sites (1 site/2.5km^2), compared to areas of Production from relatively 'natural' environments, which make up a large part of the GKLaWAC RAP area (with registered cultural site density at an average of 1 site/34km^2). In other words, there is a greater likelihood that archaeological surveys have been undertaken and cultural sites have been found and registered where the land is more intensively used. This also shows that there are noticeably fewer registered cultural sites in land that is currently protected (917 sites), than in the total amount of land characterised as Production or Intensive use areas (1781 sites), where most people live and where infrastructure is most pronounced. Again, the pattern of registered sites reflects the greater intensity of archaeological surveys undertaken for Cultural Heritage Management in areas undergoing infrastructure construction. This pattern is of relevance to the impact of fires on cultural sites because areas of denser human population, higher-density infrastructure and more managed vegetation cover are also the areas with greater fire management (and therefore fewer wildfires) (see also Chapter 5, Figure 5.4).

The pattern is also relevant to the impact of fires on cultural sites because it suggests that parts of the GKLaWAC RAP area that have lower-density human populations and limited infrastructure have received less survey coverage in Cultural Heritage Management assessments. The current VAHR records of registered sites within these areas are limited, and our knowledge of the presence of cultural sites within these areas is also limited. These areas may also be more prone to bushfires, due to their greater natural vegetation and lower readiness for fire management.

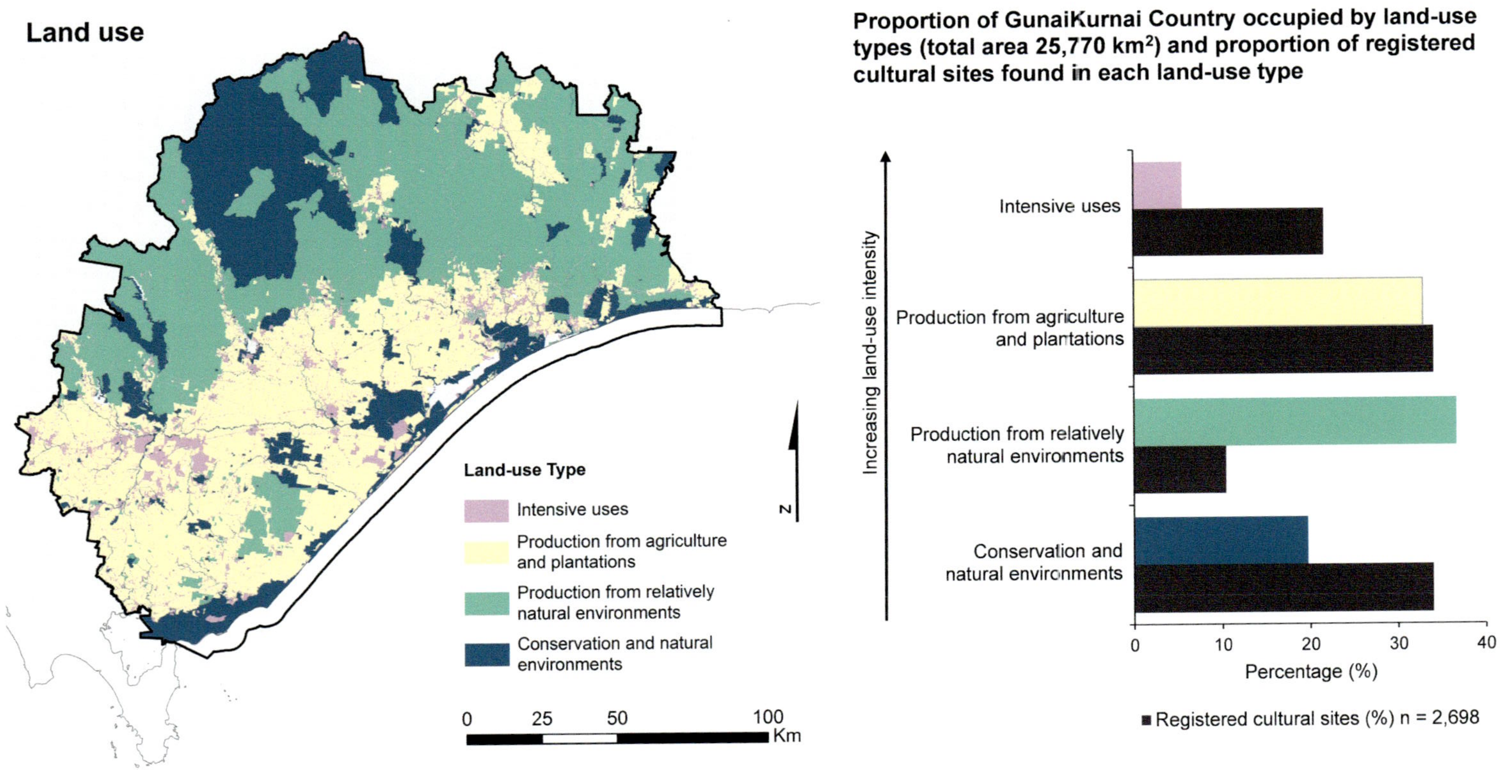

Figure 6.9. Left: Extent of each land-use type in the GKLaWAC RAP area. Right: Proportion of land in the GKLaWAC RAP area occupied by each land-use type (colours are based on the map and legend to the left) and proportion of cultural sites registered in each land-use type (dark grey bars). Land-use types are presented on the right, from bottom to top, in order of increasing land-use intensity. These data on land-use types were sourced from the Australian Bureau of Agricultural and Resource Economics and Sciences (ABARES: https://www.awe.gov.au/abares/aclump/land-use/alum-classification).

The research presented in this chapter investigated correlations between the location of archaeological sites registered in the VAHR, the surface geology of the landscape, the current vegetation cover and current land use. The results indicate that registered archaeological sites are proportionally under-represented in areas of Eucalypt Forest—Wet Sclerophyll and areas currently used for 'production from relatively natural environments'. Sites are proportionally over-represented in areas of cleared land and land designated as 'Intensive uses' or 'Conservation and natural environments'. Investigation of cultural heritage places in areas that appear to be proportionally under-represented in the VAHR records could help determine the degree to which this latter pattern results from cultural site locational choices made by the Old Ancestors, versus patterns caused by definitional preferences and mandated survey strategies enshrined in the legislation governing Cultural Heritage Management assessments. Investigating areas shown to be under-represented in this assessment of archaeological site locations across the landscape would also help develop a more representative understanding of the range of landscape locations in which the different types of cultural sites occur, providing more reliable and balanced information on the relationships between the locations of cultural places and bushfire-prone landscapes.

Chapter 7

The Impacts of Fire on Stone Artefacts

Jerome Mialanes, Bruno David, Joanna Fresløv and Russell Mullett

Stone artefact scatters are the most common type of archaeological site in Australia generally, and in GunaiKurnai Country in Victoria's Gippsland region specifically (see Chapter 6). Many of these lie on the surface, while others are buried at various depths below ground. Stone artefacts can occur across the landscape as individual finds or as dense assemblages numbering up in the thousands. In some cases, the artefacts are easily identifiable, but in the wildfire-prone landscapes of GunaiKurnai Country, stones that have been fractured by fire can be numerous and widespread. Here we present details of how fires can impact rocks, sometimes creating 'geofacts', fractured stones that resemble artefacts, and how to tell them apart. The ability of cultural heritage surveyors and site managers to be able to differentiate between the two has been stressed many times by the GunaiKurnai Land and Waters Aboriginal Corporation, representing the Traditional Owners of GunaiKurnai Country; there is a heightened concern given the large numbers of cultural heritage surveys taking place across much of the State of Victoria in the wake of the increased frequency of wildfires in recent years.

IMPACT OF FIRES ON STONE ARTEFACTS AND STONE ARTEFACT SCATTERS

Stone artefacts are found in many places across GunaiKurnai Country. Some were flaked with a hammerstone, creating artefacts such as flakes and cores. Some were ground by rubbing against another rock, creating artefacts such as ground-edge axes and grinding stones. Others yet were carried across the landscape and dropped on the ground without their shapes being modified by flaking or grinding. These latter artefacts are called 'manuports', meaning 'carried by hand', after the Latin words '*manus*', hand and '*portare*', to carry. These terms are used in archaeological and cultural heritage management surveys all over the world.

Recognising when a stone has been made into an artefact is sometimes difficult, especially when its shape has changed through natural forces such as 'insolation weathering' caused by repeated daytime warming and night-time cooling, or wildfires. The information below describes how fire can change stone, and outlines how to tell the difference between a rock that has been affected by fire versus one flaked by people.

Stone is often unintentionally burnt by wildfires or campfires. Uncontrolled heating of stone is called 'heat alteration'. However, people can also heat stone in controlled fires (or in heated sand beneath a fire) on purpose, to improve its flaking qualities. Such controlled heating of stone is called 'heat treatment'. Heat treatment is done by slowly heating stone and then slowly cooling it down. To get the best results, for most fine-grained, silica-rich raw materials such as cherts and flints, it takes between seven hours and two to three days to reach the maximum temperature (which must be held for less than one hour) and to then slowly cool down the stone (Schmidt *et al.* 2016). Sedimentary silica-rich rocks such as chert, silcrete and quartzite benefit from heat-treatment. In fine-grained silica-rich rocks such as chert, heat closes the tiny spaces between the grains and heals defects within the stone. At a microstructural level, new bonds are created, improving how the rock is held together. The rock becomes more homogeneous, fracturing more predictably and requiring less force for flake detachment when tools are made (Schmidt *et al.* 2012). This contrasts with quartz, for which heating makes it fracture into smaller pieces. Burnt quartz cannot easily be identified in the field because fluid inclusions >5μm in diameter would need to be observed under >1000× magnification (Driscoll and Menuge 2011: 2259).

THE EFFECTS OF WILDFIRES ON ROCKS AND STONE ARTEFACTS

The greater the intensity of a fire, the more it can damage stone artefacts (Buenger 2003), and the larger the surface area of an artefact, the more damage it is likely to sustain from fire exposure. Damage will also occur more rapidly where a greater surface area is exposed. Smaller artefacts are less likely to be damaged (Deal 2012: 110; Mercieca 2000: 44; Mercieca and Hiscock 2008: 2636). Wildfires will usually cause greater damage to stone artefacts lying on the ground than when they are buried, as the flames affect the stone directly. Buried stone artefacts are generally protected from wildfires by soil cover and by partially combusted fuel (Buenger 2003: 202, 205). However, the longer a fire remains in one location, the higher the chance it will damage buried artefacts. Depending on soil conditions, temperatures up to 200 °C can be reached down to a depth of 10cm (Aldeias *et al.* 2016). For archaeological sites where sediment has built up slowly, such depths would typically be hundreds or thousands of years old. Deeper and older artefacts will not usually be affected by wildfires, unless slow-burning tree roots or organic-rich peats burn below ground.

When heat is applied to stone in an uncontrolled way, it visually affects the stone in a number of ways. These are *colour change*, *crazing*, *potlidding* (creating *potlids* and *potlid scars*), *deep surface cracking* and *heat fracturing*. Stone that has been burnt can also have residues stuck to it from the surrounding vegetation. These features appear at different stages of heat application (time under heat, maximum temperature reached, and speed of heating/cooling).

Colour change. Colour change can be one of the easiest ways to tell if stone has been heated, but only if there is good knowledge of the range of raw materials available in the area and their original colours. Unfortunately, this is rarely the case in GunaiKurnai Country, a problem compounded by the great variety of raw materials in many regions and the great variety of colours of some raw material types from individual sources (e.g., the Moondarra silcrete quarry; the Glenmaggie coarse silcrete source; and many chert source locations). Knowing the original colour of raw materials is important. When the original, unburnt range of colours is not known, it can be difficult to tell if its colour has changed through burning (Henry *et al.* 2003: 16).

Each type of rock (e.g., basalt, quartzite, quartz, silcrete, chert, etc.), and each source of a given type of rock, has its own distinctive chemistry, mineralogy and structure. This means that each can react to fire in slightly different ways. For instance, Fort Hood chert (a type of chert found in the USA) changes colour by the time it reaches 200 °C, whereas Pecos chert (another kind of American chert) does not show any signs of change until 900 °C is reached, at which stage minor colour changes occur. At that temperature, the Fort Hood chert will have crumbled to pieces (Table 7.1).

Table 7.1. Thermal effects on a range of stone raw materials.

<table>
<tr><th>Temperature (°C)</th><th>Fort Hood Chert (Buenger 2003)</th><th>Pecos Chert (Buenger 2003)</th><th>Quartz (Ryan 2012)</th><th>Quartzite (Moody 1976)</th><th>Silcrete (Mercieca and Hiscock 2008)</th><th>Limestone (Chakrabarti et al. 1995)</th><th>Ground stone (Ryan 2012)</th><th>Basalt (Ryan 2012)</th></tr>
<tr><td>100</td><td></td><td></td><td></td><td></td><td></td><td></td><td rowspan="2"></td><td rowspan="2">No physical changes</td></tr>
<tr><td>200</td><td rowspan="3">Colour change</td><td></td><td></td><td rowspan="3">Peak colour change from 250 °C followed by blotching till 450 °C</td><td></td><td rowspan="2">Progressive reddening (if limestone contains iron)</td></tr>
<tr><td>300</td><td></td><td></td><td></td><td rowspan="5">Smudging, organic materials present begin to diminish-pollen</td><td rowspan="8">Spalling, fracturing</td></tr>
<tr><td>400</td><td></td><td></td><td></td><td></td></tr>
<tr><td>500</td><td>Potlid fractures, spalling</td><td></td><td></td><td rowspan="2">No physical changes</td><td></td><td></td></tr>
<tr><td>600</td><td>Extensive damage</td><td></td><td rowspan="5">Blackening, thermal expansion, crystalline structure change (>573 °C)</td><td rowspan="5">Cracks, Fracturing</td><td rowspan="3">Reduced strength, Disintegration if thermal shock</td></tr>
<tr><td>700</td><td>Fracturing + deterioration</td><td></td><td></td></tr>
<tr><td>800</td><td>Fracturing</td><td></td><td></td><td>Organic material diminished-animal proteins</td></tr>
<tr><td>900</td><td>Crumbling</td><td>Colour change</td><td></td><td></td><td></td></tr>
<tr><td>1000</td><td></td><td></td><td></td><td></td><td></td></tr>
</table>

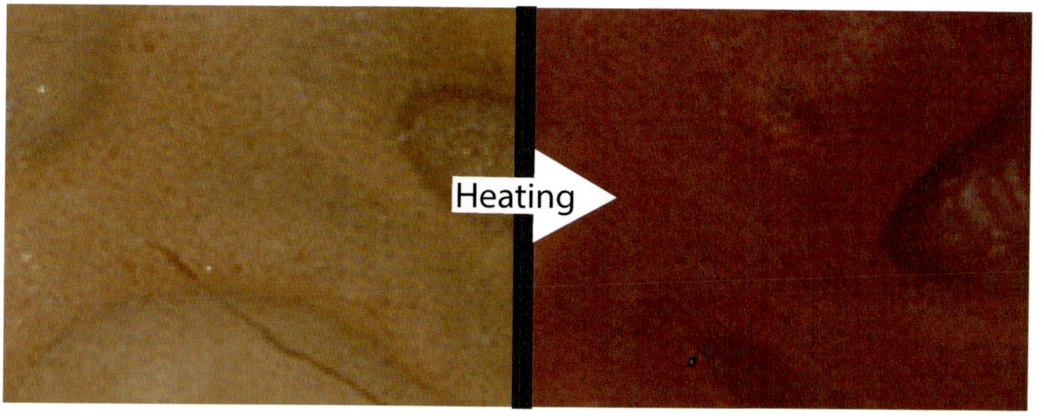

Figure 7.1. Example of heat-induced colour change on silcrete (after Schmidt, Nash et al. 2017: Figure 2).

In iron-rich rock, heat causes the oxidation of the iron mineral and changes the original colour of the material to a reddish hue (Figure 7.1). Not all rocks contain iron and therefore some will not change colours in this way when heated.

Similarly, the chemical composition and minerals in limestone will determine whether its colour will change when heated. Some limestones change from a yellowish-beige colour to grey at 400 °C. Limestones containing iron oxides can turn to red when heated to 100 °C, getting redder until 300 °C, and then change from red to grey at 400 °C (González-Gómez *et al.* 2015: 188).

Crazing. Crazing shows up as fine, twisted networks of thin and shallow surface cracks on rock surfaces (Figure 7.2).

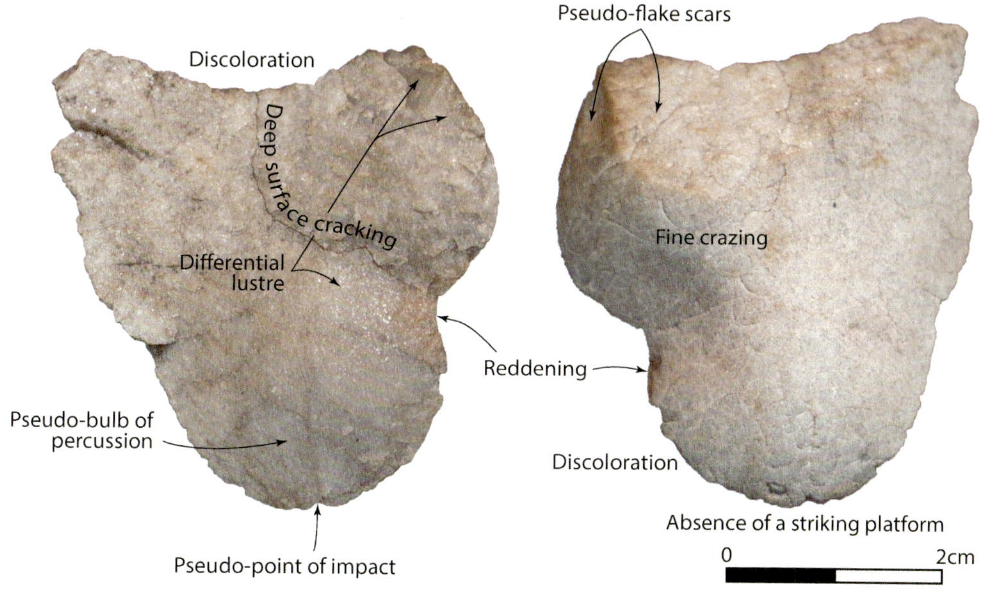

Figure 7.2. Heat-fractured rock missing a striking platform and that contains a location on its edge that resembles a point of force application (after Neubauer 2018: 685, used here with permission of the author).

Deep surface cracking and heat fracturing. Thermal shock from rapid temperature change damages a rock's crystal structure. Surface cracks become progressively deeper with increasing cycles of thermal shock (Figure 7.3; see also Fiers *et al.* 2021). These changes are accompanied by a loss of weight, increased porosity (air spaces in the rock) and an overall reduction in structural resistance (strength) (Guler *et al.* 2021).

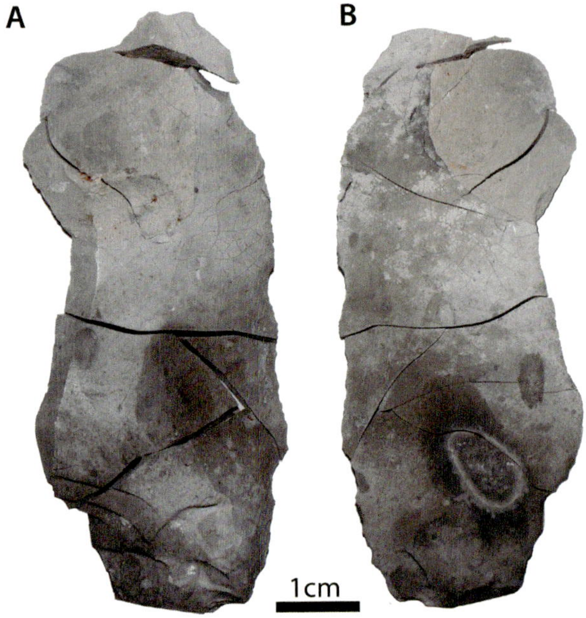

Figure 7.3. Deep surface cracks on a flint artefact following heating to temperatures between 600–650 °C (photographs: Géraldine Fiers, used here with permission of the author).

Potlids and potlid scars. Potlids are a type of rounded (in plan view), flat-topped and half-domed (when seen from the side) pieces of stone that pop off the surface of a rock as a result of thermal fracturing (Figure 7.4). They are caused by 'differential heating and pressure release probably due to steam buildup in areas of the material that has impurities or high moisture content' (Buenger 2003: 26). Potlids can range in diameter from less than 1mm to more than 6cm (Patterson 1995: 74). The mark that is left on a rock after a potlid detaches is called a 'potlid scar'. While potlids are technically not flakes, larger specimens can be used as tools or cores (Gould 1976: 142). On limestone, potlids do not occur unless siliceous inclusions such as chert are present (Pagoulatos 2005: 291).

Gloss. The presence of gloss on a rock surface can indicate that it has been heated. Gloss, or lustre, is the way light reflects on the stone, creating an oily appearance.

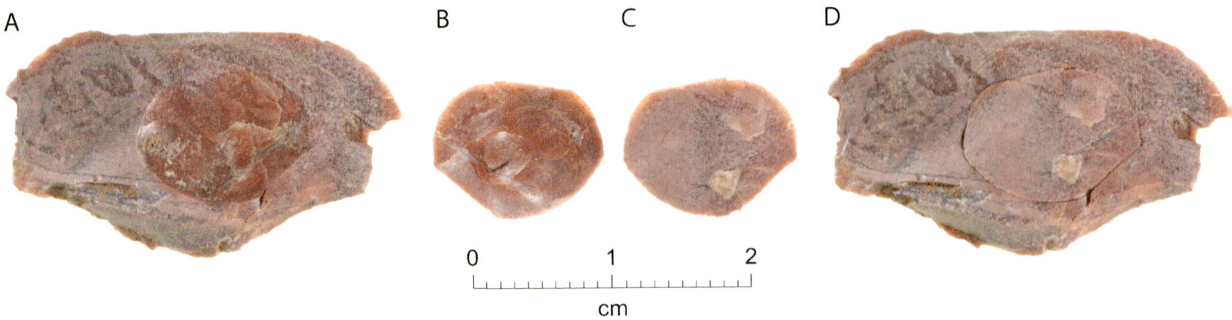

Figure 7.4. Potlid scar (A) and potlid (B: ventral and C: dorsal views), and after refitting (D) on chert artefact from Caution Bay, Papua New Guinea (photographs: Steve Morton).

DIFFERENTIATING BETWEEN ROCKS THAT WERE HEAT-TREATED BY PEOPLE VERSUS THOSE THAT WERE MODIFIED BY WILDFIRES

When uncontrolled heat fracturing occurs, such as in a wildfire, thermally fractured rocks can sometimes look like stone artefacts. Several features can help confirm if a fracture was caused by fire rather than by flaking (Figure 7.2). Heat-fractured rocks do not have striking platforms; a striking platform is a morphological feature of a flake caused by the impact of a hammerstone. Striking platforms are also absent in rocks detached by exfoliation, where repeated changes in temperature cause the surface layer(s) of rocks to detach in thin pieces that can otherwise resemble flakes.

In frost-shattered rock, broken pieces can sometimes have flat surfaces similar to striking platforms, but they lack points of force application (the spot where a rock is hit to make a flake).

Although heat fracturing can happen in any kind of fire, people sometimes put rocks in controlled fires to break them through thermal shock, especially at quarries to break large rocks into more manageable size (Akerman 1979; Binford and O'Connell 1984; Elkin 1948; Gregg and Grybush 1976: 190; Pétrequin and Pétrequin 2002; Stout 2002; Thomson 2005). Thermal shock would break rocks along existing lines of weakness within the rock. It would be difficult to differentiate rocks broken by the heat of wildfires from those purposefully broken by people using controlled fires (Florek 1989; Tibbett 2005).

Crenated fractures. Another characteristic of heat-fractured rocks is the presence of crenated (uneven or wavy) fractures (Figure 7.5). This type of fracture is the result of rapid cooling (Neubauer 2018: 683; Purdy 1971: 52)

Gloss contrast. The presence of differential lustre, or gloss contrast, on a stone artefact can prove macroscopically that a rock has been heat-treated (Schmidt, Spinelli Sanchez and Kind 2017). The presence of gloss contrast on a rock can indicate that one part of the rock (the outside) broke before heating (the rougher surface), but another (e.g., a flake scar) was made after heating (the smoother, glossy surface) (Schmidt *et al.* 2018: 593) (Figure 7.6).

The glossiness of a rock surface is partially determined by the roughness of the surface (Schmidt, Spinelli Sanchez and Kind 2017: 18). On silcrete, gloss does not appear with controlled heat treatment, but a rock that was heated before being flaked will

Figure 7.5. Crenated fracture on a limestone fragment, Millukmungee Shelter 1, a site near the junction of the Snowy and Buchan Rivers in GunaiKurnai Country (photograph: Steve Morton).

have its outer (pre-heating) surface relatively coarse, and its inner (post-heating) flake scar smooth (Schmidt 2019; Schmidt *et al.* 2019) (Figure 7.6). This contrast in gloss and/or texture cannot be achieved by any other means (e.g., wildfire or proximity to hearth, each of which affects the whole artefact) (Schmidt and Hiscock 2019: 80). Identifying heat treatment is difficult, particularly when diagnostic features have been removed from the heated artefact by subsequent flaking (Schmidt *et al.* 2018: 593).

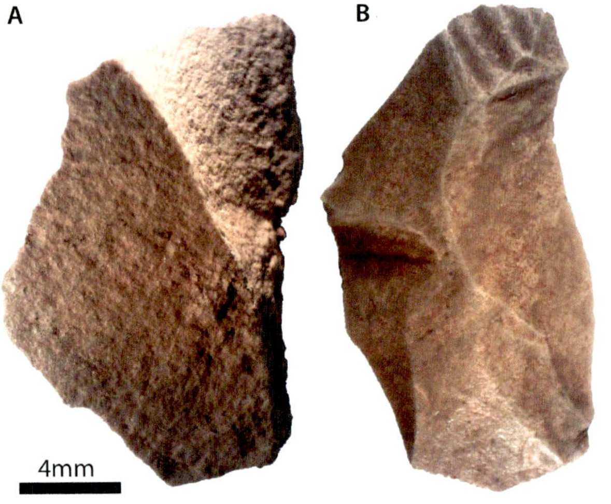

Figure 7.6. Gloss-contrast on two silcrete artefacts from Puritjarra, central Australia. Note the difference in flake scar texture between unheated Flake A and heat-treated Flake B (after Schmidt and Hiscock 2020: 3).

An absence of gloss coupled with the presence of potlid scars and crazing on a complete flake surface indicates fire damage *after* the artefact was made (Deal 2012: 100). The presence of potlid scars on the ventral surface of a flake that has no gloss or texture contrast is a sure sign of post-depositional heating (i.e., that the stone was heated after the flake was made, such as by a wildfire).

LIMESTONE

Limestone is common in some parts of GunaiKurnai Country, especially near Buchan. Limestone that has broken naturally can be particularly difficult to tell apart from limestone that has been flaked. For this reason, we summarise key ways in which limestone can break naturally in pieces that resemble artefacts.

Limestone, and rocks generally, can break naturally in a number of ways:

Thermal fatigue (e.g., insolation/onion-skin weathering). This is a type of weathering common in warm areas. As the sun shines on a rock during the day, it causes it to expand. During the night, the rock contracts due to the colder temperature. This repetitive expansion/contraction causes the rock to break 'along pre-existing lines of weakness, e.g., crystal boundaries' (Hall and Thorn 2014: 7) and for small pieces of surface rock to flake off.

Thermal shock. This is 'a single stress event whereby sudden (large) changes in temperature produce fracture because of the resulting stresses exceeding the capacity of the rock to adjust other than through instantaneous failure' (Hall and Thorn 2014: 1). Limestone will break when subjected to rapid changes in temperature. Wildfires are the main cause of thermal shock, and rocks in campfires can also shatter through thermal shock.

Chemical weathering. Rainwater, which is slightly acidic, can slowly dissolve rock and can change the chemistry of limestone to form pavements. This occurs on the surface and along the joints and bedding planes of limestone.

Physical impact in high-energy environments. In riverine and coastal environments, water-flow such as surf, wave-action, currents, and flooding, can have sufficient power to shatter limestone.

Gravity. Rock will break through gravity during mass-events such as landslides, earthquakes, colluvial erosion or roof-fall.

In summary, we highlight five major points about how limestone can be affected by wildfires (heat alteration):

1. Not all limestones react similarly to heat, as their compositions are source-specific.
2. The chemical composition and minerals in limestone will determine whether its colour will change when heated. Some limestones change from a yellowish-beige colour to grey at 400 °C. Limestones containing iron oxides can turn to red when heated to 100 °C, getting redder until 300 °C, and then change from red to grey at 400 °C (González-Gómez *et al.* 2015: 188).
3. Some limestones can fracture at temperatures between 200–500 °C and disintegrate at temperatures above 600 °C (Zhang *et al.* 2017).
4. Limestones containing chert inclusions can show evidence of potlidding when subjected to sudden and extreme changes in temperature (e.g., wildfire) (Pagoulatos 2005: 291).
5. Because limestone especially can naturally break in ways that mimic flaked stone artefacts, we provide below a key to aid differentiate between the two (Table 7.2). By following each step by checking for the presence/absence of specific attributes, flaked stone artefacts can be separated from naturally broken rocks that in various ways resemble artefacts ('geofacts').

Table 7.3 provides a key to identify hand-held ('unipolar') flaked artefacts and general flake fracture types. By following each step, the different kinds of flaked artefacts recognised in archaeology can be identified.

CONCLUSION

Stone artefacts are of fundamental importance both for the GunaiKurnai Traditional Owners of Gippsland in southeastern Australia, and for archaeologists. A key reason for this is that, being made of rock, they endure in the landscape. As wildfires rage across all of Gippsland's landforms, some site and artefact types, such as culturally modified trees (see Chapter 10), are being increasingly destroyed, to the point that their numbers are rapidly diminishing. In this context of both increasing landscape fire regimes coupled with forest clearance, agricultural expansions and increasing urban and related development spread, stone artefact scatters are quickly becoming the most visible type of cultural heritage site in the landscape (albeit along parts of the coast, shell 'middens' remain prominent) (see Chapter 14). Their endurance in the landscape is thus of heightened importance to avoid the erasure of the physical signs of ancestral Aboriginal presences as time unfolds, as, given the state of things, neither wildfires nor infrastructure developments are expected to slow down, let alone cease, in any part of GunaiKurnai Country in the foreseeable future. The ability to identify, and with this to differentiate, stone artefacts from 'geofacts' resembling but not made by people, is thus of

Table 7.2. Key to assist in the differentiation between naturally fractured stone and whole artefacts flaked by hand-held ('unipolar') percussion. By following the key from the start, stone flakes versus stones with ventral-like surfaces broken by wildfires and other 'natural' causes can be differentiated.

Key	Criteria	Go to
1	The point of force application is present	2
	The point of force application is absent	3
2	**It is a flaked artefact**	
3	The location of the point of force application can be identified	4
	The location of the point of force application cannot be identified	5
4	The exterior platform angle is less than 90°	6
	The exterior platform angle is more than 90°	5
5	**It is a geofact**	
6	There is a prominent bulb of percussion on the ventral surface	7
	There is no prominent bulb of percussion on the ventral surface	8
7	Flake scars are present on the dorsal surface	9
	Flake scars are absent from the dorsal surface	10
8	Ripple marks are present on the ventral surface	7
	Ripple marks are absent from the ventral surface	5
9	Dorsal flake scars are parallel to the medial axis	2
	Dorsal flake scars are not parallel to the medial axis	11
10	The striking platform is convex	5
	The striking platform is flat	2
11	The flake presents one or more retouched margins	2
	The flake does not present any retouched margins	12
12	Use-wear is present on at least one margin	13
	Use-wear is not present	5
13	It is a used (but not necessarily flaked) stone artefact	

importance, both for the recording and management of sites, and for their recognition as enduring physical expressions of the Old Ancestors in fundamentally *cultural* landscapes. The recognition of stone artefacts in the landscape is a recognition of the presence of the Old Ancestors.

Table 7.3. Key to unipolar flaked artefact identification (after Hiscock 2007: figure 11.2).

Key	Criteria	Go to
1	Negative or positive flake scars are present	2
	Negative or positive flake scars are absent	3
2	**It is a flaked artefact**	4
3	**It is not a flaked artefact**	
4	An unambiguous ventral surface is clearly visible	5
	The ventral surface is missing or obscured	6
5	One or more positive flake scars are present	7
	Positive flake scars are absent	8
6	**It is a flaked piece**	
7	**It is a flake**	9, 12
8	**It is a core**	
9	Flake scars were produced after the last ventral surface was created	10
	Flake scars were not produced after the last ventral surface was created	11
10	**It is a retouched flake**	
11	**It is an unretouched flake**	
12	All margins are intact: **It is a complete flake**	
	One or more margins are absent	13
	No margins are present: **It is a flake fragment (other)**	
13	Only the distal margin is present: **It is a distal flake fragment**	
	Only the lateral margins are present: **It is a medial flake fragment**	
	Only the proximal margin is present: **It is a proximal flake fragment**	
	Only the distal margin is present: **It is a distal flake fragment**	
	The flake is broken along its longitudinal axis: **it is a longitudinal flake fragment**	
	Only one lateral margin is present: **It is a marginal flake fragment**	

Chapter 8

The Impacts of Fires on Rock Art Sites and Ochre

Jillian Huntley and Courtney Webster

Rock art can be created by the addition of pigments (drawn or painted images, or stencils and prints of objects), or the subtraction of bedrock (engravings, including scratched images) (see also Chapter 6; e.g., Dibden 2019; O'Connor *et al.* 2013 in relation to scratched rock art). Only three rock art sites have been documented from GunaiKurnai Country, probably reflecting biases in the location of the surveys undertaken during cultural heritage management (see Chapter 12). The three previously recorded GunaiKurnai rock art sites are situated in the southern foothills of the High Country, in association with outcropping bedrock (Table 6.1; Figures 6.2, 6.3). Wildfires, particularly in the 1930s and 2000s as illustrated in Figure 5.3, were geographically concentrated in the foothills and High Country, in areas of remnant forest preserved in National Parks and other 'public land' associated with the ancient bedrocks of the Great Dividing Range and Australian Alps (Agriculture Victoria 2022). One of these three known GunaiKurnai rock art sites is a rock shelter with a single white painting near the entrance of Cloggs Cave (for details of Cloggs Cave, see David *et al.* 2021; Delannoy *et al.* 2020). The second is Mitchell River Shelter, a large rock shelter in Mitchell River National Park. According to its recording form in the VAHR, 'graffiti is present on the walls of the shelter—one set of initials has the date "1965". One set of lines does not appear to be initials and may be of Aboriginal origin'. It was recorded in 1986 and has not been investigated since, and it is not clear whether the set of apparently engraved lines is indeed 'rock art'. The third site is New Guinea II, a limestone cave surrounded by dense temperate wet sclerophyll forest and rainforest along the western bank of the Snowy River. The cave is richly decorated with rare finger-flutings and a few other engravings on its walls and ceiling, in internal parts of the cave where the indirect sunlight is subdued and where light does not reach. To date, only one GunaiKurnai art site has been studied in detail—the white painted image on the wall of the rock shelter at the entrance to Cloggs Cave (Figure 8.1; Webster 2021). It is worth noting that rumours of other rock art sites in the GKLaWAC RAP area have been reported by local residents in rural areas, and are planned to be investigated by GKLaWAC.

IMPACTS OF FIRE ON ROCK ART SITES

Rock art is made on bedrock outcrops: rock shelter and cave walls/ceilings, or horizontal stone platforms. Rock art is therefore susceptible to the impacts of heat described in this volume for rock shelters, tool stone and stone technologies (see Chapters 6 and 7). As noted in these chapters, the effects of heat will vary depending on the host rock. In Victoria, sandstone, granite and limestone are the most common bedrock types on which rock art was produced (see Chapter 6). The impacts of heat on drawings, paintings, stencils and prints will also vary depending on the properties of the pigments used. Rock art Earth pigments (commonly called 'ochres' in Australia) are typically made from siltstones, carbonates, clays, calcites, iron-rich lenses/gravels and sulphide deposits. These collectively occur in hues of purple, red, orange,

Figure 8.1. Cloggs Cave painting (after Webster 2021) (photograph: Courtney Webster).

yellow, black, green, white and even blue, while wood and bone carbon (even mineral coal) have been used as black pigment (Bonneau *et al.* 2011; Chalmin and Huntley 2018; Wallis *et al.* 2016). Heat impacts on ochres are discussed in detail below.

There is potential for information about the lifeways of the Old Ancestors to be incorporated into every stage of rock art production, leaving a record of the materials and tools they used, the steps they went through to prepare pigment/paint, or to remove stone in the case of engravings. As with tool stone and shell (see Chapters 7 and 11, respectively), rock art can be affected by uncontrolled heat from wildfires, but the deliberate application of fire is also known to have been used to change the properties of ochres as part of art production in many parts of Australia and beyond (Chalmin *et al.* 2008; Pomiès *et al.* 1999; Salomon *et al.* 2012, 2015). Here we therefore discuss the effects of heat from uncontrolled fires on rock art, with heat treatment (deliberate burning) of ochres addressed in the next section.

Intense, severe wildfires cause more damage to cultural sites than cooler cultural burns. Many contributing factors in fire behaviour are known (Chapter 2), with risks such as fuel build up and the position of sites in the landscape applying equally to rock art as to all other cultural heritage. For instance, an engraved rock platform at the top of a ridgeline near the apex of a slope that fire can run up may be at greater risk of more frequent and intense burns. Similarly, fuel build up from vegetation within and in front of rock shelters, and at the edges of, or on, horizontal engravings can prolong heat impacts and increase fire temperatures. For example, studies show the dire consequences of thermal shock where prolonged and/or intense heat is created in rock art sites by emplaced site infrastructure such as boardwalks and railings being

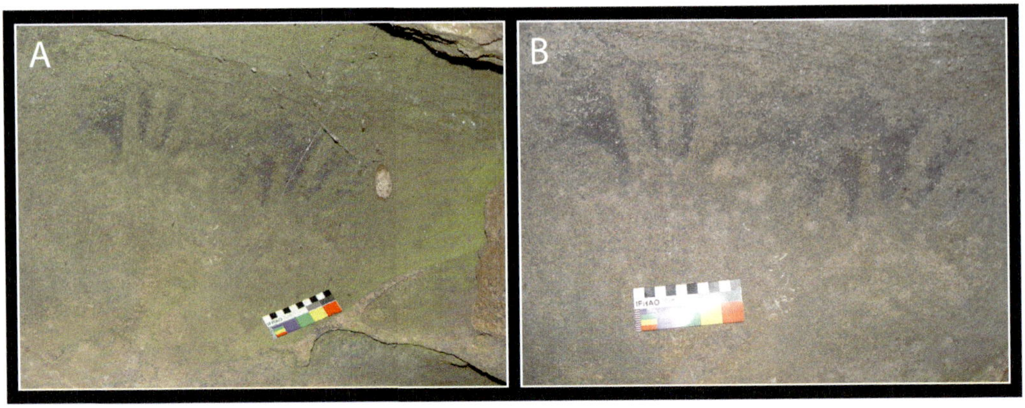

Figure 8.2. Lichens and other microflora on the art panel at Breakfast Creek 45 site, on the Woronora Plateau, Sydney Basin, New South Wales. A: In late 2002. B: The same panel free from microflora following a controlled burn of vegetation in the Sydney Catchment Metropolitan Special Area, August 2004 (after Ford 2006) (photographs: Jillian Huntley).

set alight (Lambert and Welsh 2011; Taçon and Harding 2019). The removal of vegetation, leaf litter and fallen timber to control risks are common tasks in many Indigenous Ranger and Caring for Country programmes (e.g., Cole and Wallis 2019; Marshall *et al.* 2020), and controlled/prescribed (cool) burns are a common and effective mitigation measure in cultural burning regimes (Chapter 2; Freeman *et al.* 2021; Sefton 2011). Prescribed burns can also have additional beneficial effects. For example, cool burns can inhibit other agents of rock art deterioration such as the growth of lichens and algae (microflora) (Ford 2006; Figure 8.2).

Just as thermal shock produces spall flakes in locations where heat has reached bedrock surfaces (Lacanette *et al.* 2017), so too rock art will spall from mechanical stress as different minerals within the same pigment shrink and swell at different rates from daily, seasonal and extreme temperature and humidity variations. Thermal shock from wildfires can result in extreme rock/pigment spalling, as minerals in rocks are generally poor heat conductors. For example, rock surfaces exposed to the sun will heat at a different rate to the underlying rock or other parts of the rock that are not exposed. Different temperatures in different parts of the rock can cause the outer surface to become detached from the underlying rock (flaking), a process known as 'exfoliation' (Hoerlé 2006; Tratebas *et al.* 2004; Twidale 1968: 137). In the case of wildfires, heat can be applied to rock surfaces causing the rapid swelling of susceptible minerals. In this way also, different rock art pigments have different thermal properties. For instance, red and white pigments respond differently to changes in temperature because of their different thermal (and other physical) properties, which is a factor in pigment loss, leading to mechanical stress between pigment layers, and mechanical stress between pigments and rock surfaces (Hall *et al.* 2007; Huntley *et al.* 2014, 2021).

For paintings, drawings, stencils and prints, it is important to understand the properties of pigments as they influence preservation, acting as a surface modifier, changing reflectance, light transmissivity, thermal and moisture properties on the rock face (Sumner *et al.* 2009: 242). The smaller particle size and conductivity of iron minerals in red ochres makes them physically stable when heated, even to extreme temperatures of *c.* 900 °C (Khan and Chhibber 2021). White ochre clays and carbonates including calcites, however, have large particles, sheet-like structures and are porous. White ochres are highly susceptible to changes in heat and humidity, with large shrink and swell capacity making them some of the most mechanically unstable pigments. Added to this, calcite, other carbonates, and to a lesser extent, silicate minerals (clays) are soluble, with a tendency for the pore spaces created by their movement to be filled by weathering products such as salts (e.g., gypsum). This chemical instability can further exacerbate flaking and pigment loss (Clarke and North 1991: 92; Huntley *et al.* 2014, 2021).

In 2020–2021, at GKLaWAC's request, the white paint used to make the rock art in the rock shelter outside Cloggs Cave was investigated. The motif was made of a quartz-rich, mixed mineral clay, likely obtained from a nearby vein in the limestone bedrock (Figure 8.3). Layers of natural mineral coatings—the calcium-oxalate minerals weddellite/whewellite—were found across the rock art panel and have helped stabilise the white paints. That said, geological salts (sodium sulphate, magnesium sulphate, sodium carbonate, calcium chloride and gypsum) were found across the rock art panel, as well as mineralised limestone surfaces with a darker rind compared to fresh spall. In combination with these salts, there are microfractures along bedding planes in the finely laminated limestone bedrock, and the art panel in the Cloggs Cave rock shelter is slowly flaking (Figure 8.3; Webster 2021). The flaking of the rock art panel at the Cloggs Cave rock shelter could be exacerbated and accelerated by fire, though the nature of the landscape and low fuel loads make the likelihood of fire reaching the rock art panel very low. In early 2020, the Gippsland fires burned through the landscape that contains this site, but the art panel has not been visually affected in any obvious way (it is not known exactly how close to the rock wall, and painting, the wildfires got).

Colour change in rock art pigments can also be caused by heat exposure from fires, though it is more common and widespread as a result of sun exposure (insolation). In Kalkadoon Country (northwest central Queensland), research suggests that yellow rock art has turned red over time from repeated wetting and drying cycles of pigments and heat from sunlight in a region where summer air temperatures frequently reach more than 50 °C (Cook *et al.* 1990). Colour change happens because dehydration catalyses a transition of the mineral goethite (generally yellow) to haematite (generally red). A secondary impact from wildfires can be increasing insolation, at least in the short term, as vegetation cover surrounding rock art sites is removed, exposing pigments and surfaces to more sunlight. As well as flaking from mechanical stress, increased sunlight changes the conditions on the rock face and can facilitate the growth of microflora, such as lichens, across rock art panels (Huntley 2019; Huntley *et al.* 2018) (Figure 8.4).

Both painted and engraved art are more susceptible to secondary impacts after wildfire(s) through enhanced chemical weathering processes, even where thermal shock is not obvious (Twidale 1968). This is because the intense swelling and subsequent shrinking of rock art pigments, and the bedrock surfaces that host art, creates pore spaces and small cracks (microfractures, similar to the crenated surfaces from uncontrolled heat on stone tools; see Chapter 7). These tiny spaces can be filled by water and eventually mineral deposits, weakening the bonds between stone/pigment, creating further mechanical stress (Ford 2006; Huntley *et al.* 2021; Twidale 1968). Therefore, the true impacts of wildfires on rock art are unlikely to be seen for years, even decades, after an extreme fire event, although sometimes more seasonal effects can be imminent. This was the case in Ngunnawal Country (Namadgi National Park), where initial thermal shock to the Rendezvous Creek rock art

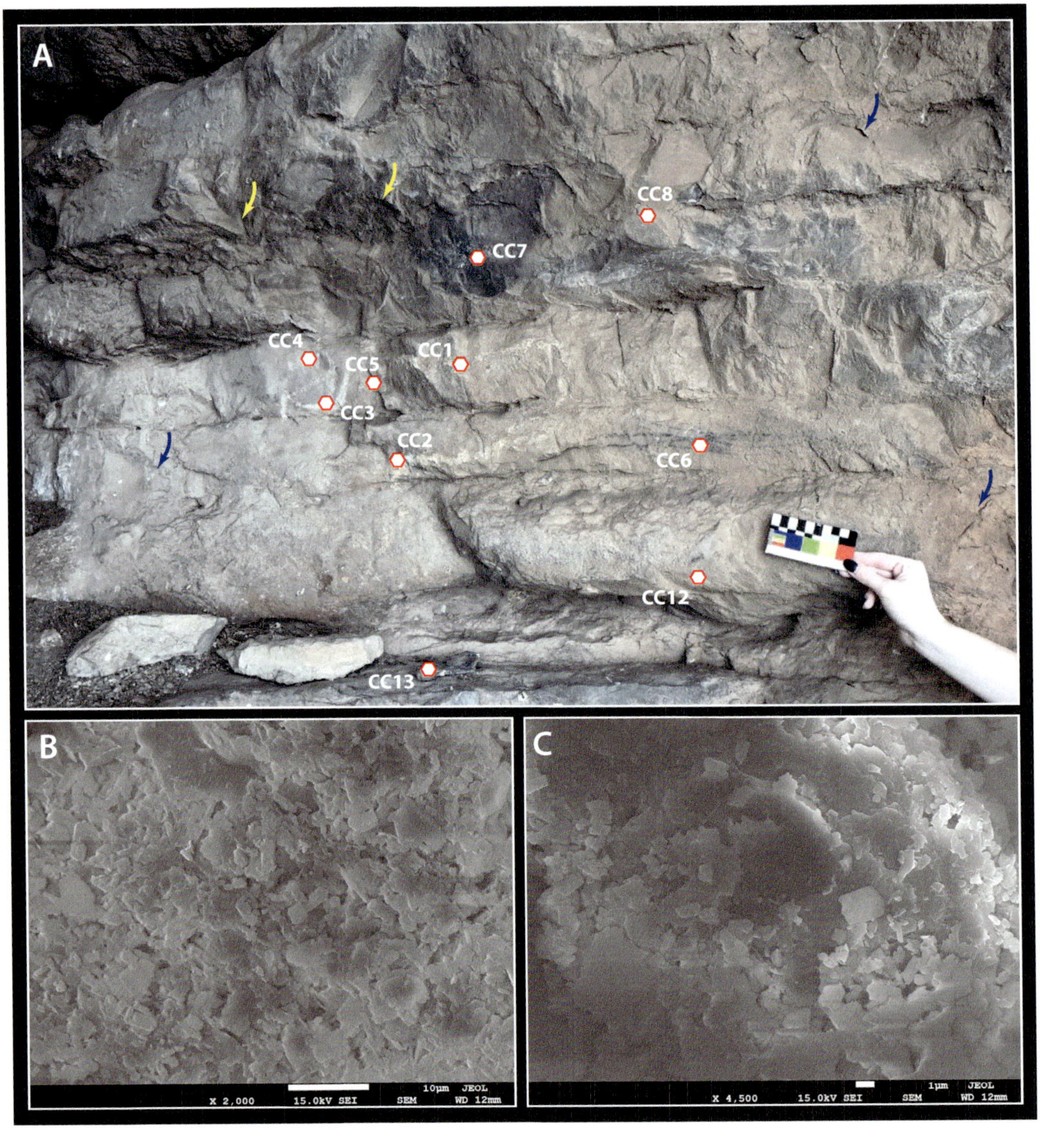

Figure 8.3. The rock art motif located in the Cloggs Cave rock shelter, showing both new (blue arrows) and old (yellow arrows) flaking and spalling of the rock surface, as well as the locations of small samples taken for laboratory analysis (top: photograph taken by Jillian Huntley). Scanning Electron Microscopy images showing the clay-mica pigment (bottom left: sample location CC2) and calcium oxalate (weddellite) crystals (bottom right: sample location CC1), both taken at 2000× magnification (after Webster 2021).

site (containing a wooden viewing platform) in the devastating 2003 Australian Capital Territory (ACT) wildfires (also known as the 'Canberra' wildfires) was followed by a later, secondary block fall and spalling of rock surfaces after the first winter frosts (Wallis *et al.* 2003; Watchman 2004a, 2004b).

Figure 8.4. Detail of lichen growth following increased sunlight on the art panel from the removal of eucalypt forest on the slope behind the sandstone boulder in Kabi Kabi Country (Sunshine Coast, Queensland) (after Huntley *et al.* 2018) (photographs: Jillian Huntley).

Finally, while both bedrock surfaces and rock art pigments can change colour as a result of heat exposure (Cook *et al.* 1990; Lacanette *et al.* 2017), the numerous accounts of 'smoke blackening' from flames touching stone surfaces and/or campfires having been lit inside caves/rock shelters (e.g., Raper 2003) commonly reported in site records, cultural heritage management survey reports, and academic papers, are likely grossly overstated. Over the years, a number of us (including JH, Bruno David and Jean-Jacques Delannoy, personal communication 2021) have been taken to dozens of rock art sites where archaeologists and other researchers have assumed that black colouring on rock shelter and cave walls and ceilings are the result of soot and/or charring from fires. In almost every case this has been a misidentification of microflora (generally lichens) and the deposition of minerals (particularly iron-rich accretions) in solution—that is, minerals deposited by water running over or evaporating at the edge of stone ceilings and walls. In other cases, dark coloured dust particles, either from ash or increased sediment erosion, can cover rock art following wildfires. In most cases this is temporary, though dark particles trapped in mineral accretions like silica skins in sandstone rock shelters can permanently obscure art (these may be datable, generating minimum ages) (Huntley and Officer 2016; Tratebas *et al.* 2004; Watchman 1993).

IMPACTS OF FIRE ON OCHRES AND OCHRE SOURCES

Ochres are mineral pigments—colour-producing stones, clays and soils. The term 'ochre' has mostly been used by archaeologists to describe red, iron-rich, Earth minerals. Iron-rich ochres are generally red but also commonly purple, yellow, and orange, and sometimes appear black, brown, and even green (Rifkin 2011). Aboriginal peoples across Australia also use the term 'ochre' to describe a variety of Earth minerals, including clays, that may be coloured grey, white, and even blue (e.g., Chalmin and Huntley 2018; Huntley 2021; Wallis *et al.* 2016). Apart from producing rock art and body adornment, ochres have many functions including medicines for wounds and the eradication of parasites (Mphuthi 2013; Plomley 1966; Velo 1984) and as a barrier to protect people against the sun, cold and insects (Backhouse 1843; Brian 1979; Davies 1846; Plomley 1966; Sagona 1994: 23). It is unlikely that ochre use was ever just prosaic in Aboriginal societies, because landscapes are intertwined with the activities of the Ancestral Beings who created them (Blundell and Woolagoodja 2005; Head *et al.* 2005: 257; Scadding *et al.* 2015; Vinnicombe 1997). In other words, ochre sources are a part of cultural landscapes, storied places of significance to communities.

Ochre sources may be in primary contexts such as conglomerate pebbles, veins and seams in bedrock. Ochres can also be found in exposed sediments and in secondary source locations like loose cobbles in creek/river gravels, and nodules in alluvial/colluvial sediments. Secondary ochre sources have originally come from the primary source locations, and have then been transported across the landscape by natural (geomorphic) processes. Ochres are almost always mixed minerals, unique to the place where they were formed by geological weathering. Scientific analyses can be used to work out where they have come from or understand the ways they have been prepared (such as by deliberate heat treatment). Because the differences between ochres are not always obvious, scientific analyses may be the only way to tell apart ochres from difference sources or find out if heat treatment or other processing has occurred (Huntley 2021). For example, the famous Wilgie Mia and Little Wilgie ochre quarries in Wadjari Country (Weld Range, Western Australia) are in the same geological seam, but have distinctive chemical signatures, so researchers could test hand stencils in the area to find out which ochre source was used to make them (Scadding *et al.* 2015).

GunaiKurnai Country encompasses a number of geologies with suitable raw materials for tool stone and pigments as both (primary) outcrops and bedrock seams, and secondary sources in watercourses and alluvium/colluvium (transported stones and sediments) (Huntley 2021). As Figure 7.3 shows, the large size of sedimentary and regolith (alluvium and colluvium) distributions that occur in GunaiKurnai Country should have a number of mineralogically distinct quarries within them, and the discrete nature of igneous and metamorphic geologies means there is good potential for successful geochemical sourcing (Dayet *et al.* 2016; MacDonald *et al.* 2013; Scadding *et al.* 2015; Smith *et al.* 1998). Importantly, as with stone tools, the primary distinction between 'raw' and heat-affected ochres is a loss of water molecules.

Figure 8.5. Examples of excavated ochre pieces with use-worn surfaces, from northern Australia. A, B: Facetted surfaces with microstriations, Nawarla Gabarnmang. D: Facetted surfaces with microstriations, site JSARN-113/23. C: Edge-rounding, Nawarla Gabarnmang. E: Edge-rounding, smoothing and microstriations, Nawarla Gabarnmang (photos by Steve Morton, courtesy of Bruno David).

So, while mineral transitions and associated colour changes may occur as a result of heat, the trace element chemistry of the ochres remains the same, meaning that scientific sourcing studies will likely be successful irrespective of the impacts of fires (Huntley 2021).

Pieces of ochre can be a feature of artefact scatters, shell middens and buried archaeological deposits, or they can be found by themselves within sites (Chapter 6). As detailed in Table 6.1, there is a total of 2214 such registered sites in GunaiKurnai Country that have the potential to contain archaeological ochres—those used by the Old Ancestors. We know that ochres have been used where the Earth minerals are found outside their natural geological occurrences (manuports), and/or where evidence of wear marks such as flaking, grinding or rubbing is preserved (Figure 8.5). In addition, ochre sources can be recorded as quarry locations where there is preserved evidence for material extraction, such as through the preservation of flaking, pounding, gouging, grinding and/or scratching marks (Figure 8.6). Ochres are also a common feature of burials in many parts of Aboriginal Australia, especially secondary interments where bones were often ochred (e.g., Bowler *et al.* 2003). There are 23 recorded burial sites in GunaiKurnai Country, but none are known to have been ochred (Table 6.2).

Pieces of ochre associated with surface sites are more likely to be badly impacted by uncontrolled heat from fires. As with stone tools and rock shelters, fire impacts are more likely to include uneven colour changes (such as from yellow to red or red to mulberry/purple on the edges of ochre nodules), crazing (many micro-fractures), or abrupt/explosive fracture as water rapidly evaporates (heat shatter). On the other hand, heat treatment generally increases the colour saturation of roasted ochres and can create a greasy luster, especially if the ochre matrix has a lot of quartz. Just like understanding where ochres have come from in the landscape, it can be difficult to tell if a piece of ochre has undergone heat treatment by just looking at it, but scientific analyses can be used to make such determinations (Chalmin *et al.* 2008; Godfrey-Smith and Ilani 2004; Legodi and de Waal 2007; MacDonald *et al.* 2019; Marcaida *et al.* 2017; Opuchovic and Kareiva 2015; Pomiès *et al.* 1999; Salomon *et al.* 2012, 2015).

Ochres buried in archaeological deposits can undergo colour change from chemical interactions, and many archaeologists suspect that red ochres are over-represented in the archaeological record as a result (e.g., Wadley 2009). However, as fires heat soils at their surface, the amount of red ochre, and often lack of white ochre, found in archaeological deposits probably results from iron (red ochre) being more chemically inert than clay and calcite minerals (white ochre). Especially in acidic sediments like sandstone rock shelters, chemical weathering is more likely to slowly erode white ochre and change yellow ochres to red.

In summary, pieces of ochre within artefact scatters, shell middens and buried archaeological deposits, have the potential to be present in thousands of currently registered GunaiKurnai sites, and untold numbers of unrecorded and unregistered sites. Surface, or near-surface ochre artefacts, especially in open context where large fuel loads of vegetation are present, could be affected by fire causing thermal fracture, crazing and colour change (including charring). As fires heat deposits at their surface, it is very unlikely that ochres present in deeply stratified archaeological deposits could be impacted by uncontrolled heat—though tree roots and the like can slowly smoulder below the ground, and the emplacement of infrastructure such as boardwalks may provide fuel sources for fire, increasing the depth of thermal penetration.

Very few rock art sites are known from GunaiKurnai Country, with the only two definite ones being in dense wet forest near the western bank of the Snowy River, and in an elevated rockshelter with very little burnable vegetation nearby. Currently neither is likely to be directly impacted by fire. As with all heritage places, the materiality of GunaiKurnai rock art sites significantly determines to what extent, and in which ways, they are likely to be impacted by fire. Situated in the southern foothills of the Great Dividing Range and Australian Alps, the area near or around the two previously recorded rock art sites and a third, possible rock art site have experienced intense wildfires, particularly in the 1930s and 2000s, and in the case of Cloggs Cave, 2020. No fire impacts have been recorded to the rock art of GunaiKurnai Country, although none of these three sites has been inspected with fire damage in mind

Figure 8.6. Ochre quarries. A–C: From Banjima Country in the eastern Hamersley Range of the Pilbara region, Western Australia, with evidence of percussion flaking. D, E: From Wunambal Gaambera Country in the northwest Kimberley, with evidence of scraping (after Wallis *et al.* 2016 (left) and Huntley *et al.* 2015 (right)).

following a wildfire. Furthermore, the true impacts of wildfires on rock art are unlikely to be seen for years, even decades, after an extreme fire event; the location of rock art in deep rockshelters/caves, within riparian zones and remnant damp forests means the likelihood of intense, severe wildfires directly reaching the rock art panels is low. A key finding from previous research across Australia and beyond continues to promote the removal of vegetation, leaf litter and fallen timber and controlled/prescribed (cool) burns adjacent to rock walls that have rock art, as a means of minimizing the effects of fires that could otherwise burn onto, or very close to, the rock art sites.

Chapter 9

The Impact of Fires on Bone

Matthew McDowell

IMPACTS OF FIRE ON BONE

Fire is a fundamental feature of the Australian landscape. Uncontrolled wildfires can reduce large tracts of vegetation to ash in just a few hours, but in the hands of someone with experience in cultural burning, fire can be used to reinvigorate vegetation patches to promote the retention of animal and plant biodiversity. In this chapter, I discuss how bone responds to being burned at different temperatures and over varying durations. Understanding these burning patterns on bones can help interpret the archaeological record and interrogate what diverse signatures of burning mean in terms of a site's fire history, both in Australia and beyond.

The origin of burnt bone in archaeological sites has long been debated (e.g., Brain 1983). Bone recovered from archaeological sites may have been burnt through a number of processes, most of which relate to human activity. Bone might be burnt:

- During food preparation.
- During site clean-up and rubbish disposal.
- To make sure it is disposed of by fire in a culturally appropriate way to ensure its smell does not attract spirits, as is common in many Australian Aboriginal communities.
- As fuel.
- Inadvertently by lighting a fire over partially or completely buried bones.
- By landscape fire, either wildfire or cultural burning of Country.

As people are directly responsible for all these processes except wildfires, it is often assumed that burnt bone provides evidence of people and what they did. However, the temperatures at which a bone was burnt leaves characteristic features, including bone colour, uniformity of damage, fractures, and altered surface textures that can help understand how it was burnt (Nicholson 1993).

Food preparation. During food preparation, be it in an oven (such as an earth oven) or on an open fire, bone is largely protected from direct exposure to intense heat by muscle tissue (Solari *et al.* 2015). Therefore, apart from a little charring of exposed ends (David 1990), cooked bone shows little macroscopic evidence of modification compared with bone exposed directly to higher temperatures. However, if cooked bone is examined using scanning electron microscopy and a range of X-ray diffraction techniques, it will readily show morphological changes indicative of low temperature heating of the collagen fibrils (Koon *et al.* 2003; Solari *et al.* 2015).

Fires in GunaiKurnai Country (Archaeopress 2023): 109–113

Rubbish disposal, cultural disposal in campfires, and bone as fuel. When exposed to intense heat (e.g., when burnt for cultural reasons after a meal, during rubbish disposal or to fuel a fire), bone can undergo several stages of chemical and physical change. This begins with dehydration, followed by degradation of organic material, degradation of inorganic material, and finally the oxidisation of hydroxyapatite (calcination) (Castillo *et al.* 2013; Nicholson 1993).

The most obvious, and perhaps most subjective, physical change associated with burnt bone is its colour. Colour has long been used to identify archaeological bones modified by heat. The relationship between colour and temperature has been extensively researched and tested experimentally, with colour changes largely indicating the sequential decomposition of the organic and inorganic components as heat increases (Correia 1997). When fresh bone is exposed to heat (Figure 9.1), it follows a general progression from brown through black, then grey, to blue or blue-grey-white to white with increasing temperature (Table 9.1). Unburnt bone is typically cream to yellow in colour, but when exposed to low temperature (100–150 °C) it can develop brown spots. When heated to 180–200 °C, bone takes on a deep brown colour. When heated to about 300 °C, it blackens ('carbonises'). When heated to 400–650 °C, it turns grey and begins to change chemically. When bone is heated to more than 700 °C, it undergoes inversion (the loss of bone carbonate) at which point it may turn blue, blue-grey, blue-grey-white or white. At this point it undergoes 'calcination', oxidising and dramatically changing its structure (Asmussen 2009; Baby 1954; Bennett 1999; Binford 1963; Buikstra and Swegle 1989; Fernández-Jalvo *et al.* 2018; Gilchrist and Mytum 1986; Hermann and Bennett 1999; Knight 1985; Mays 1998; McCutcheon 1992; McKinley 2000; Marques *et al.* 2018; Nicholson 1993; Shipman *et al.* 1984; Stahl 1996; Stiner 1991; Stiner *et al.* 1995) (Figure 9.1). However, despite these well-studied temperature-related colour changes, the relationship between colour and temperature can vary if bone was contaminated with sediment or metal oxides when burned. Therefore, the colour of a bone should be treated as providing a general indicator of the maximum temperature to which a bone has been exposed, rather than a specific or precise temperature (Shipman *et al.* 1984). Furthermore, if bone is heated for a long time at a low temperature, it may mimic the characteristics of bone that has been burnt over a short time at a higher temperature (Asmussen 2009). However, it is unlikely that bone will be calcined in landscape fires, where ground-level temperatures in Australian conditions do not generally reach beyond c. 300 °C, a temperature that can result in bone carbonisation. The rare exception is if a bone burned over a prolonged period of time inside, or leaning against, a burning log, or in prolonged-burning peat deposits.

Changes to bone during an animal's life is commonly referred to as 'bone biogenesis', and after an animal's death 'bone diagenesis'. Several colour classification systems have been developed to describe how bone changes through burning, but other factors such as bone size, shape, and the thickness of the cortical bone may also influence bone diagenesis through fire (e.g., Correia 1997; Devlin and Herrmann 2008; Shipman *et al.* 1984; Symes *et al.* 2008).

Table 9.1. Typical temperatures required for burnt bone to reach a range of colours.

Burning Stage	Bone Colour	Temperature Reached (°C)
1	Cream-yellow	Unburnt
2	Spotted brown	100–150
3	Brown	180–200
4	Black	300
5	Grey	400–650
6	White	≥700

Figure 9.1. Bones burned at different temperatures or durations can take on different colours. The most common colour-temperatures are shown here (all animal bones showed here are from archaeological excavations in GunaiKurnai Country). A: Unburnt bone (from Wangangarra 1 excavation square M10 XU29A). B: Browned bone, 180–200 °C (from Millukmungee Shelter 1 square O5 XU10). C: Charred bone, 200–300 °C (from Millukmungee Shelter 1 square O5 XU10). D: Greyed bone, 400–650 °C (from Millukmungee Shelter 1 square O5 XU11). E: Blue-greyed bone, 500–650 °C (from Millukmungee Shelter 1 square P8 XU24). F: Calcined bone, ≥700 °C (Millukmungee Shelter 1 square O5 XU10). G) Black and blue, calcined (from Millukmungee Shelter 1 square O5 XU10). This bone was probably burnt in a campfire. Different parts of the bone were exposed to different temperatures, causing different levels of bone modification. The white surface of the bone has been calcined, the blue edge of the calcined bone was heated to a slightly lower temperature, and the blackened/charred remainder was heated to a lower temperature again (photos by Steve Morton, montage by Bruno David).

Some take a simple and archaeologically easier to apply approach, qualitatively describing the colour associated with a temperature range, such as we have done above. However, others (e.g., Castillo *et al.* 2013; David 1990; Shipman *et al.* 1984) use the Munsell colour classification system to accurately describe the colour of burnt bone at a given temperature and duration of burning, to limit inconsistencies between researchers when describing bone colours.

In addition to colour change, burnt bone also experiences microscopic physical changes. Castillo *et al.* (2013) recorded histological changes in the organic and inorganic bone matrix when subjected to heat. When heated between 100 and 300 °C, bone develops an undulating, 'glassy' surface with occasional longitudinal fractures. Between 300 and 400 °C, collagen denatures and the bone surface may be covered with a peeling, bubbly layer of char. Between 400 and 600 °C, hydroxyapatite is disassociated and collagen is thermally destroyed, making the surface look granular. Extensive polygonal cracking around articular surfaces can also occur at such temperatures (Nicholson 1993). Bones burnt between 700 and 800 °C are characterized by a loss of homogeneity in the bone's crystalline structure, and bones exposed to temperatures over 900 °C exhibit a granular structure with amorphous crystal clusters and a complete loss of osseous structure. Interestingly, bone exposed to temperatures over 1000 °C regains a smooth compact surface made up of micro-crystals, and all surface irregularities disappear (Castillo *et al.* 2013).

Incidental burning. When fires are built over bone buried by sediments, the burning signature of de-fleshed and dry bones may be overwritten, replacing the original taphonomic signature. Experimental results indicate that the majority of de-fleshed bones directly exposed to a hearth-fire becomes calcined, while those buried between 1 and 15cm beneath a hearth-fire become carbonised (Stiner *et al.* 1995). This is consistent with the small mammal bones recovered from Cloggs Cave (McDowell *et al.* 2022). Experiments conducted by Bennett (1999) and Stiner *et al.* (1995) found that the surface colour on the majority of buried fresh bones exhibited 'continuous colour' resulting from burning uniformity. However, when older bones experience similar treatment, colour changes are observed to be less extreme or muted (Bennett 1999; Buikstra and Swegle 1989: 255; Stiner *et al.* 1995).

LANDSCAPE FIRES: WILDFIRES AND CULTURAL BURNING

The effects that landscape fires (wildfires and cultural fires) can have on bone depend on the bone itself (features such as its size, shape, and the thickness of the cortical bone), the temperature and the duration of exposure, and the fuel and oxygen supply (Bennett 1999; Correia 1997; Stiner *et al.* 1995; Thompson 2004). In extreme conditions the base of wildfire flames can reach up to 1100 °C, exceeding the temperature needed to calcine bone. However, the visible flame tips rarely exceed 300 °C (Wotton *et al.* 2011), and the hottest flames (and thus hottest flame bases) are usually located some distance above ground level, and so will not be in contact with bone lying on the ground. Unlike a hearth which is fed fuel to maintain high temperature combustion for several hours, a wildfire must keep moving to acquire fuel and oxygen. Therefore, while a wildfire may briefly generate the temperature required to calcine bone, it will rarely generate that temperature in one place long enough to cause calcination. Furthermore, the bones of animals that die in wildfires are protected by muscle tissue. Consequently, bones are rarely exposed to wildfires directly. Only animal carcasses that have been skeletonised prior to a wildfire will be fully exposed to the heat it generates. Usually, such bones from natural deaths will occur as either isolated bones or the bones from a single isolated animal in the landscape (unlike cultural bone accumulations, which often consist of multiple animal remains from meals or other cultural activities).

Therefore, while wildfires and landscape-scale cultural fires have the potential to briefly subject bones to temperatures achieved in camp/cooking fires, the duration of exposure is generally too short to have the same effect as bone burnt in or immediately under fireplaces.

DISCUSSION

Differentiating between bones burnt by landscape fires such as wildfires and bones burnt in dedicated fireplaces by people often simply comes down to how long the bones were burnt for, along with the temperatures reached. Wildfires do not typically stay in one place long enough to burn bone beyond carbonisation. Furthermore, if the bones were part of a carcass, they would mimic features associated with cooked bone. By comparison, a hearth may be kept alight for several hours resulting in bone calcination. Even if the temperature was too low to calcine bones, if bones are left in a fire long enough, they can reach the same degree of modification as if they were burnt at very high temperatures.

Chapter 10

The Impacts of Fire on Culturally Modified Trees

Joanna Fresløv, Russell Mullett and Bruno David

In Aboriginal society, as elsewhere, trees are not just botanical species. They are a part of Country—the living world formed by the ancestral beings at the beginning of time, and that continues to connect people and place as kin. Trees are expressions of culture, and how trees are experienced and known reflect the philosophies of that culture.

The plant ecologist Ross Hynes and social anthropologist Athol Chase employed the term 'domiculture' to describe how researchers could think about relationships between people and plants across cultures (Hynes and Chase 1982). Instead of trying to cram Indigenous ways of being into the alternative pigeonholes of 'agriculturalists' versus 'hunter-gatherers', they argued that it would be much more meaningful to try to understand and describe local Indigenous ways of being on their own terms. When it comes to communicating how people interact with plants and animals, a better way than to use the terms 'agriculture' and 'hunting and gathering', they argued, is to try to understand precisely how people interact with organisms and their communities. How are plants meaningful in culture, for example? How do people look after them? How do they engage with them in everyday life? How do plants fit in to their philosophies of life and world views? This way, a much better understanding would be gained of the culture of trees and other plants, and why looking after them in certain ways and by certain people matter.

Trees, groves and more generally whole plant communities are often of great cultural importance, and this is so probably of all peoples around the world. Individual trees may mark the places where ancestors reside(d), such as at Poromoi Tamu on the bank of Utiti Creek, in the southern lowlands of Papua New Guinea. Poromoi Tamu is an ancestral story-place of the Himaiyu clan among the Rumu (Figure 10.1A). Here, on the eastern bank of Utiti Creek opposite Are Creek, five coconut trees grow. They are the descendants of the coconut trees that had been planted generations earlier, when a small village thrived. Today all traces of the village have gone, except for the coconut trees that mark where the ancestral village was, and where the events retold in oral traditions took place (for details, see David *et al.* 2012; for another story-tree among the Keipte Kuyumen, 70km to the northwest, see David *et al.* 2015). The trees are markers of story-places that connect people to place, to the ancestors, to what is meaningful in the landscape, and to each other.

Around the world, trees can also house named spirits, or they can be named spirit-beings in and of themselves (e.g., Figure 10.1B). In the Kimberley region of northwest Australia, some prominent trees were carved open to create cavities large enough to imprison people overnight; oral histories and written documents survive of some of the atrocities that took place where Aboriginal people were chained onto trees in early colonial times, or where they were used as holding cells (e.g., Bohemia and McGregor 1995; Figure 10.2). They are historical landmarks, emotional places, markers of injustices that continue to reverberate

Figure 10.1. Important cultural trees of the Himayiu clan, Rumu cultural group, southern lowlands of Papua New Guinea. A: The coconut trees that mark the place of the Himaiyu clan's ancestral village place of Poromoi Tamu, eastern bank of Utiti Creek. The trees have grown from coconuts that fell from previous generations of palm trees thought to have begun in the 18th or 19th centuries. The current coconut trees are visual markers of Himaiyu clan social and landscape (hi)stories. B: The Himaiyu clan spirit-tree Uerera (tall tree in centre of photo), in which live two spirit-men, Mumai and Ehepekai, eastern bank of Utiti Creek (photos by Bruno David).

with present-day generations, and living monuments of the colonial landscape. Elsewhere in Aboriginal southeastern Australia, trees were sometimes carved with powerful geometric designs to mark the locations of graves and communicate their kinship affiliations (e.g., Black 1941; Etheridge 1918). Across much of Australia also, trees were modified by cutting toe-holds (also called 'toe-holes') for climbing, whether to reach the upper heights of the trunks to obtain possums or 'sugar-bag', the honey from stingless bees. Small openings in the tree were expanded to reach the targeted possum or honey, and the scars on such trees can survive to this day (e.g., Morrison and Shepard 2013).

Most commonly, the modification of trees relates to the bark being cut away in sheets from the trunks to make canoes, walls or roofing for shelters, as canvases for painting, or to wrap ancestral remains. Travelling around Country and seeing such trees in the landscape is a reminder and confirmation of 'we were here', 'we *are* here', as were our ancestors way back in time onto the present. In GKLaWAC Country, culturally modified trees are in many ways *the* most visual expression of GunaiKurnai culture in the landscape, because the signs of the ancestors are there readily visible on the land. Caring for Country is caring for the trees is caring for the ancestors is caring for the people today. The Old Ancestors may be deceased, but they have not passed *away*, visibly present in the trees they interacted with or into which they have returned.

CULTURALLY MODIFIED TREES IN GUNAIKURNAI COUNTRY

Culturally scarred and shaped trees are some of the most noticeable traces of the Old People in GunaiKurnai Country. The trees line the highways, country roads and tracks, waterways, lakes, farm paddocks and even towns and housing estates (Figure 10.3). To GunaiKurnai, the First Peoples of this Country, they are more than archaeological traces of the past: they are physical and spiritual connections with the ancestors, history. They are historical documents, monuments, markers of time and place. They tell of a time when the Old People lived, travelled through and shaped the forests, cared for the land, its animals and plants, and named its landmarks. In a similar yet different vein, Greg Lehman (2000: 1) wrote poignantly of the disappearance of Palawa (Tasmanian Indigenous peoples) cultural traces in the Tasmanian landscape as the trees grow back: 'The bush—the land itself; once home to *Palawa*, is reclaiming the culture it once nurtured'. Culturally modified trees—their gradual decay, death, destruction and loss—are particularly vulnerable across Australia. As in Tasmania, time, people, fire and the elements are gradually erasing them from the land. As markers of ancestral Aboriginal presences, and as expressions of culture and living heritage, they mark the land as important '(hi)story places' of GunaiKurnai life and Country.

Much of the Gippsland landscape of GunaiKurnai Country is dominated by forest (Figures 10.4, 10.5; see Chapter 14). As Ruth Pullin describes in Chapter 4, however, the landscape the European settlers found when they first encountered GunaiKurnai Country in the mid-1800s was very different to how it is today. Forests were much more extensive, more open and park-like. Forests and woodlands were cultural artefacts shaped and managed by purposefully lit, relatively cool and relatively small ('patchwork') landscape fires that burned the undergrowth and the 'ladder fuels' that enable wildfires to climb from the forest floor to the crowns of trees. Since the mid-1800s, the open forests have been cleared to make way for farms. Colonial forest-clearance practices involved ringbarking and hot burns to clear the bush, while the

Figure 10.2. The Wyndham 'prison tree', a large baobab tree (*Andasonia gregorii*) artificially cut open to lock up Aboriginal prisoners in the 1890s and subsequent years (e.g., Anonymous 1931; Ribber 1940) (photos by Bruno David).

Figure 10.3. Culturally modified trees in public places in and near the town of Bairnsdale, East Gippsland, GunaiKurnai Country. A: Scarred Tree with bark removed from a River Red Gum, Howitt Park. B: Scarred Tree repositioned in a public part in Eastwood. C: Scarred Tree repositioned from a nearby area of GunaiKurnai Country into a roundabout at Eastwood (photos by Joanna Fresløv).

more remote forests were and continue to be logged. Plants regrew across many areas and over vast distances, but these did so as thick stands of more-or-less unmanaged trees of similar ages, with few old-growth trees remaining or cared for. Today the trees, and the forests, are largely left at the mercy of increasingly severe and expansive wildfires, the major force that shapes the forests of today (Lindenmayer *et al.* 2020).

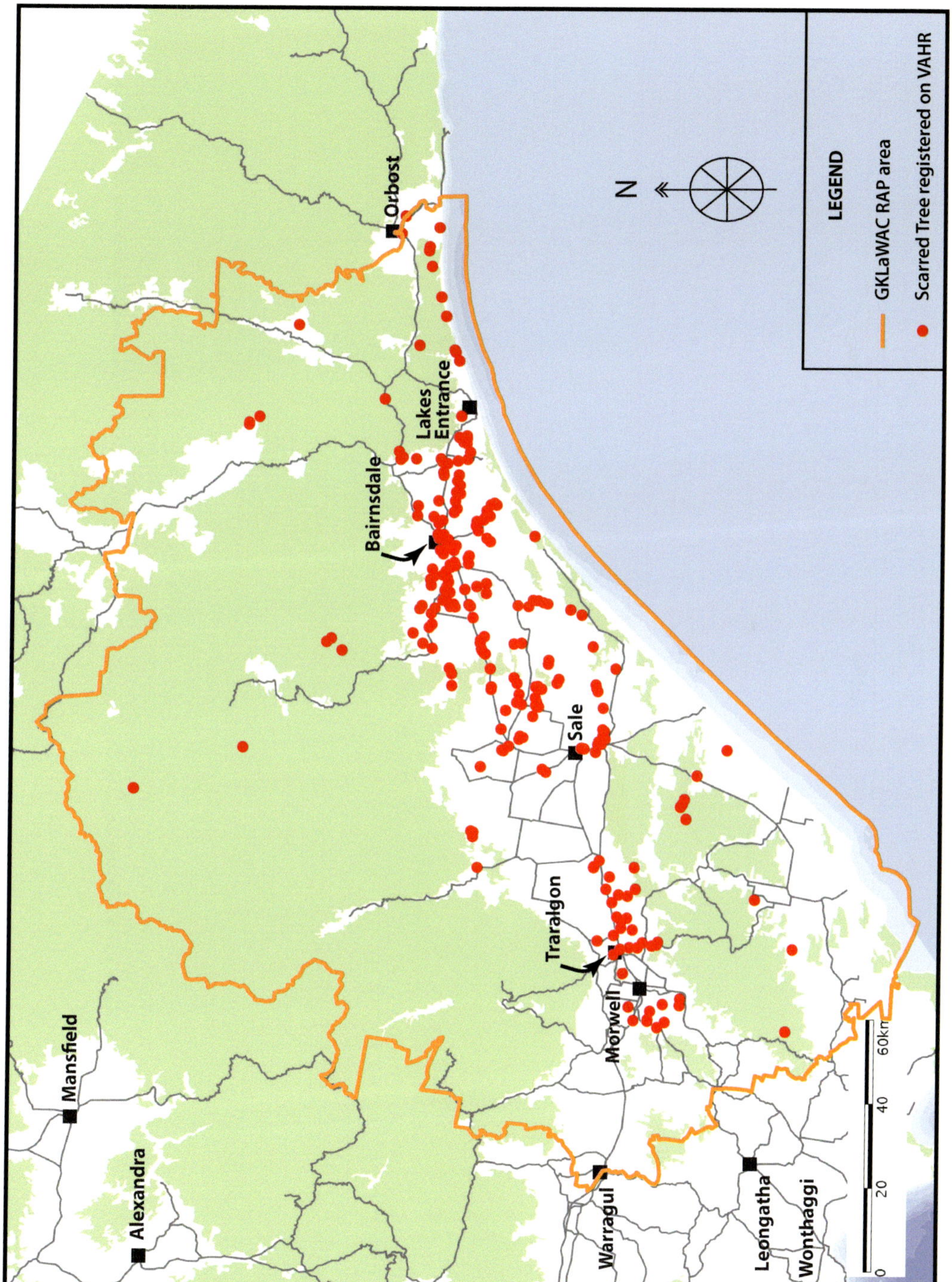

Figure 10.4. Culturally modified trees recorded in GKLaWAC Country since 1971 and registered on the Victorian Aboriginal Heritage Register (VAHR), relative to forested areas (green shading) (base layer and tree cover: from GeoVic Tree Cover; Scarred Tree data: from VAHR).

Figure 10.5. Dense forests and woodlands in the GKLaWAC RAP area that have undergone very limited to no surveying for cultural heritage sites. A: Coastal woodlands, Corringle. B: Snowy River forest east of township of Buchan. C: Dense forest near the Snowy River. D: Elder Russell Mullett undertaking cultural heritage surveys along the Mayford Track, view east toward the Dargo High Plains. E: Crystal Bay, Lake Tyers (photos: A, C, D, E by Joanna Fresløv; B by Bruno David).

Most of the known and registered culturally modified trees in the GKLaWAC RAP area are what are commonly known as 'scarred trees', and indeed 'Scarred Trees' is the registering term for culturally modified trees used in the Victorian Aboriginal Heritage Register (VAHR; see Chapter 6). In Gippsland, scars primarily resulted from the removal of bark to make canoes (Figure 10.6), shelters, containers and shields, and for the removal of strips for fibre for nets and fishing lines. Some of these Scarred Trees retain cut marks around the edge of the exposed wood from the one-handed axes used to remove the bark, but most do not. When present, the older cut marks were made with stone axes, the more recent ones beginning in the mid-19th century increasingly with metal axes. Bark was sometimes stripped to obtain grubs for food (Long 2003: 14; Smyth 1878: 207). Trees also became scarred when the bark and wood were removed to provide zig-zagged notches ('toe holds') to enable people to climb the tree to obtain possums in the higher branches. These particular scars are often associated with smoke holes at the foot of the tree, where small fires were lit to smoke out the possums in hollows in the trunks. Sometimes there are also holes higher in the tree, that enabled access to the possum nest ('drey') or 'tree hollow' (Long 2003: 13). A. W. Howitt (in Smyth 1878: 202), an early explorer and ethnographer in East Gippsland, noted that sheets of bark were used to make huts that were sometimes decorated inside with drawings or paintings:

> Mr A. W. Howitt informs me that it was the custom of the natives of Gippsland to strip a sheet of bark, bend it across the middle, and set it up like a tent, and draw figures inside with charcoal, or perhaps red-ochre (*nial*). He says he saw such an one on the Wonnangatta River, when prospecting, in 1861. He thinks the figures drawn were those of men, emus, &c.

Bark and wood were used in ceremony, including funeral practices and initiations. Bark was used to wrap the dead, to provide cord to tie the bark-wraps, and to build huts over the deceased for the mourning ceremony. After decomposition, the remains were sometimes buried in hollow trees; they are burial chambers (e.g., Howitt 1904: 459). In *jeraeil* ceremonies (male initiation ceremonies; '*jeraeil*' means 'leafy' in the GunaiKurnai language), bull-roarers made from wood played prominent roles (Howitt 1904: 509, 617; see also Gibson and Mullett 2020, 2021). So, too, did the trees and branches of the forest, and the forest itself which wrapped around the ceremony to provide privacy and secrecy (Howitt 1904: 628). Different types of wood, and the trees of the forest, played prominent roles in magic and belief systems of the GunaiKurnai, many of which would leave little trace among the forest trees (Howitt 1904: 376–377; Smyth 1878: 475). Similarly, modified trees which marked special locations, or acted as waymarkers, sometimes had their branches bent or trained into shape (Fels 2011: 272, 274; Howitt 1904: 722). As a result of their subtle modifications, often camouflaged in the midst of a forest of other trees, these and other kinds of modified trees can be difficult to find. This is doubly so as wildfires can themselves scar trees, meaning that finding the culturally modified ones requires even more intensive, and focused, survey attention. The cultural scars caused by people are structurally different and with careful scrutiny can be differentiated from the scars caused by fallen branches and wildfires (Figure 10.7).

Aboriginal Scarred Trees are trees that have physical traces of cultural modification made by the Old People in the past, rather than trees that have been scarred by natural forces (such as lightning scars or fire scars), or that have scars resulting from more recent modifications (e.g., ring-barking by early colonial settlers). The physical evidence used to determine

Figure 10.6. GunaiKurnai woman Kitty Johnson (Brabraulung clan) in her bark canoe at the Lake Tyers Mission Station, 1865 (photo courtesy of State Library of Victoria).

Figure 10.7. Examples of trees scarred by wildfires at Ewings Marsh, GKLaWAC RAP area (photos by Anita Barker).

whether a tree was modified by people, and by Aboriginal people in particular, includes the age of the tree (if it is more than *c.* 150 years old, it will pre-date the settler colonial period in GunaiKurnai Country), species, removal marks (stone axe cuts, sometimes metal-axe cut marks), the regularity of the scar shape, the depth of bark overgrowth over the wood exposed by the cut (relative to the age and health of the tree), placement of the scar above the ground, and in many cases Indigenous or local knowledge.

GUNAIKURNAI CULTURALLY MODIFIED TREES RECORDED ON THE VICTORIAN ABORIGINAL HERITAGE REGISTER

There are currently 387 known culturally modified trees in the GKLaWAC RAP area registered as 'Scarred Trees' on the VAHR ('the Register') (Figures 11.4, 11.8; see Chapter 6). Official recording of culturally modified trees began in 1971, just as the State legislation aimed at protecting Aboriginal cultural heritage places was being drafted. Half (50%) of the culturally modified trees on the Register were recorded during that decade. Most of these trees were recorded in lowland farm paddocks by honorary inspectors called 'wardens', protectors of Aboriginal heritage appointed under the *Archaeological and Aboriginal Relics Preservation Act* 1972 (now superseded). During the 1980s, site registrations resulted from surveys often in the more remote forests, with most trees recorded adjacent to dirt tracks or within a few metres of tracks in the highlands foothills and High Country. Therefore, while there are relatively good distributional data for the lowlands, those for the foothills and High Country are much more sparse and significantly biased to relatively recently disturbed locations along tracks and roads. The known distributions and densities of culturally modified trees as recorded in the Register are therefore very likely to be seriously underrepresented for the more remote parts of the foothills and High Country.

Bark-removal scars were made on a range of tree species in Gippsland's forests, with most (70%) made on River Red Gum (*Eucalyptus camaldulensis*) or Forest Red Gum (*Eucalyptus tereticornis*) (Table 10.1, Figure 10.9A). Discussing early observations of how trees were used by Aboriginal peoples in southeastern Victoria in the 1800s, Smyth (1878: lviii) mentioned that bark for canoes ('*gri*' in the GunaiKurnai language) was also obtained from Mountain Ash (*Eucalyptus regnans*). No scars on Mountain Ash have been recorded so far, but this may be because Mountain Ash is not readily amenable to scarring from bark removal, unless the inner parts of the bark are cut from the trunk.

Today there are very few large stringybark or red gum trees with the potential to have been used for canoe-bark removals left in Gippsland. Those that are large enough often do not have the appropriate strength or grain in the bark, in that defects such as burs and bore holes from insects have rendered them of insufficient smoothness and strength to remove in the one, large sheet. Following the 2003 alpine wildfires that devastated the High Country in and beyond GunaiKurnai Country, 14 areas were targeted for archaeological survey, including in forests of Mountain Ash (Fresløv *et al.* 2004). Good ground-surface visibility following the fires allowed more off-track surveying in what had been dense forest prior to the fires. A total of 18.4ha of Mountain Ash and mixed Mountain Ash forests were effectively surveyed, without finding any scarred Mountain Ash trees (Fresløv *et al.* 2004: 170).

Figure 10.8. Examples of culturally modified trees in the GKLaWAC RAP area. A: Scarred Tree from the Wonnangatta Valley. B: Scarred Tree at Knobs Reserve in eucalypt woodland. C: Toe-hold tree from lowland plains woodland, Perry Bridge. The tree was repositioned to this location c. 100m from its original growing locality after having fallen down. It was moved to here and set in a concrete footing by a local farmer to preserve it from further damage. D: Close-up of tree in C, showing location of toe-hold scars (photos: A, C, D by Joanna Fresløv; B by Jo Bell).

Table 10.1. Number of Scarred Trees in the GKLaWAC RAP area registered on the VAHR, by reported taxa.

Taxon	Number
Banksia (*Banksia integrifolia*)	1
Box sp.	4
Brown Stringybark (*Eucalyptus baxteri*)	1
Eucalypt sp.	34
Grey Box (*Eucalyptus microcarpa*)	1
Red Box (*Eucalyptus polyanthemos*)	11
Red Gum—River Red Gum (*Eucalyptus camaldulensis*) and Forest Red Gum (*Eucalyptus tereticornis*)	272
Southern Mahogany (*Eucalyptus botroides*)	11
Stringybark sp.	5
Swamp Gum (*Eucalyptus ovata*)	1
White Stringybark (*Eucalyptus globoidea*)	1
Yellow Box (*Eucalyptus melliodora*)	6
Unknown	39
Total	**387**

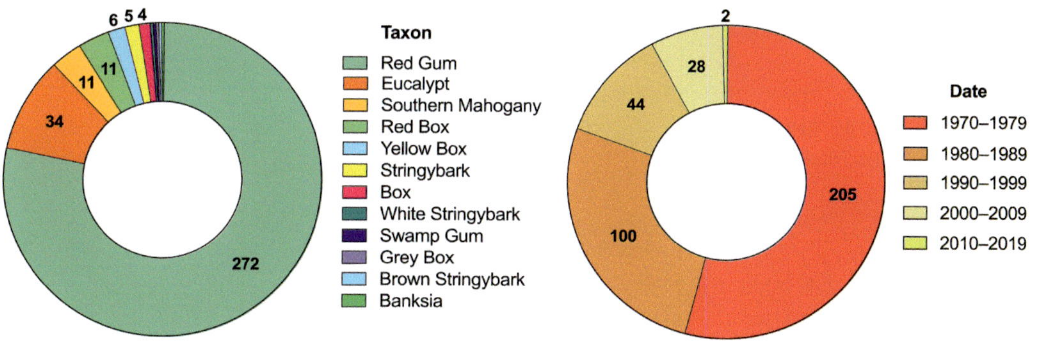

Figure 10.9. Culturally modified trees in the GKLaWAC RAP area. A: Number of Scarred Trees registered since 1971, by taxon. B: Number of Scarred Trees registered by decade (Scarred Tree data: from VAHR).

Although most bioregions of the GKLaWAC RAP area (see Chapter 14) have been subjected to cultural site surveys since the 1970s and 1980s—with these surveys having mostly targeted specific parts of the landscape, so that vast expanses of land remain archaeologically unknown—the registration of culturally modified trees has declined markedly since the early years of the Register. Indeed, only two newly found culturally modified trees have been reported in the 10 years since the beginning of the 2010s (Table 10.2, Figure 10.9B). Increasingly, after cultural heritage site legislation changed in 2007, surveys have predominantly been development-driven and focussed in those areas of the lowlands where forests and woodlands had already been cleared in the 19th century. Few extensive surveys have been carried out in the forested foothills and High Country. The 2003 post-fire surveys, in the extensively burnt forests of the High Country, recorded only one Scarred Tree in 159ha of actual surveyed area (Frésløv *et al.* 2004: 168). This Scarred Tree was in the Tea Tree Range, located in the sub-alpine zone at just over 1400m above sea level (a.s.l.). Eight other, previously recorded Scarred Trees on Mount Sarah, at 1470m a.s.l., were inspected. They were all found to have been non-Aboriginal scars made in the late 20th century during fire management activities. In the foothills of the High Country, six Scarred Trees were recorded more recently. One was found in the Wonnangatta Valley, two east of the Tambo River valley, and three in the Mitchell River valley. The forests of the foothills and High Country together cover a vast area. While cultural heritage surveys have been limited in these areas, it is very likely that many (but possibly sparsely distributed) culturally scarred trees are still present but remain unknown.

Table 10.2. Number of Scarred Trees in the GKLaWAC RAP area registered on the VAHR, by decade to 2019.

Date	Number
1970–1979	205
1980–1989	100
1990–1999	44
2000–2009	28
2010–2019	2
Unknown	8
Total	**387**

The health and survival of the 387 registered culturally modified trees in the GKLaWAC RAP area are largely unknown. Only 53 of the original 387 trees have been inspected since 2000 (Figures 10.10A, 10.10B). When originally recorded, 123 were still in good condition, while 199 were dead, dying or destroyed, and 45 were in poor health. The condition of 20 were not recorded. A review of the original site records and 2019 aerial photographs undertaken for the present study concludes that the status of most of these trees is currently unknown. Only 30 of the registered trees are confirmed to still be present in the landscape (Figures 10.10C, 10.10D, Table 10.3).

Table 10.3. Number of Scarred Trees in the GKLaWAC RAP area registered on the VAHR, by condition when they were last inspected (which in many cases = their original recording).

Last recorded condition	Number
Dead	28
Dead but standing	87
Dead, fallen	10
Dead, not *in situ*	2
Dead, burnt	1
Destroyed	70
Dying	1
Good	123
Poor	45
Unknown	20
Total	**387**

Most of the originally recorded Scarred Trees are in lowland areas that have not been subjected to repeated wildfires (Figure 10.11). The frequent and intensive wildfires of the Gippsland forests since the 1930s suggests that the almost total absence of known culturally modified trees in the High Country and extreme paucity in the foothills may not simply be a product of biased surveys. Rather, many culturally modified trees are likely to have been destroyed by wildfires, especially in recent decades when they became more frequent, intensive and expansive, as well as through commercial logging (an industry that has been pronounced across the region since the 1880s). While not all of the foothills and High Country have been logged for old-forest trees, culturally modified trees that did survive logging are in many cases unlikely to have survived the many wildfires that have raged across the landscape since the early 1900s until the 1980s, when archaeological surveys first began to be undertaken in the more remote forests (Figure 10.10).

CONCLUSIONS: TOWARDS CONSERVATION AND MANAGEMENT

The cumulative decay and destruction of culturally modified trees can occur very rapidly, especially when contrary circumstances against their survival all align. The registered Scarred Tree from the Wonnangatta Valley, in the foothills of the High Country (Figure 10.8A), has been consistently inspected since it was first recorded and in good health in 2003. Each time it has been inspected, care has been taken to rake away the accumulated 'leaf litter' or fuel from around the tree, to help protect it from wildfires. In the space of 18 years since its original recording, it has deteriorated from being a healthy living tree to a dead tree, to at-risk of falling, and, at its last inspection in 2021, to a fallen tree in very poor condition. Once on the ground, fallen culturally modified trees, like any other fallen tree in these landscapes, are very unlikely to survive the next wildfire. This is also true of the vast majority of both known and unknown culturally modified trees in the forested areas; they, too, now have a very limited anticipated survivability, given the severity of the wildfires in this region. Similar conclusions have been drawn elsewhere around the globe on this topic, such as in the Northern Hemisphere forests of Scandinavia, Canada and the USA where a combination of wildfires and logging has drastically reduced the ability of culturally modified trees to survive into the 21st century (e.g., Andersson 2005: 21, 29; Ericsson *et al.* 2003: 293; Johnson *et al.* 2022: 36; Östlund *et al.* 2005: 323; 2009: 107).

There has been very little effort made in the past to protect culturally modified trees from wildfires in GunaiKurnai Country, where the flames often reach the tree canopy and temperatures are often very high (see Chapters 2, 5). Current fire management practices include prescribed burns of moderate to cooler heat, as a calculated strategy to reduce the risks of wildfires in forests close to towns and infrastructure as 'significant assets'. More recently, members of the GunaiKurnai community have undertaken much cooler pre-emptive cultural burns (often grassfires) as a way of caring for Country steeped in traditional ways (see Chapters 1, 3). Cultural burns, and to a lesser extent prescribed burns initiated by government environmental management authorities, allow time to plan for the protection of cultural places and community heritage sites such as culturally modified trees. Recent experimentation on ways of protecting culturally modified trees has involved the wrapping of Scarred Trees with foil insulation to protect them from the flames prior to the commencement of prescribed burns and cultural burns. The areas around the trees were cleared of leaf litter to reduce fuel loads, and the wrappings removed after the prescribed and cultural fires to determine how the trees and their scars fared (Figure 10.12). Neither the scars nor the trees showed any signs of burning under their protective sleeves. But these were controlled experiments in medium to low heat that did not have flames that reached the tree canopies—wildfires are not so choosy. Nevertheless, the success of these experiments suggests that trees can to some degree be protected from less intensive fires.

The destruction of culturally modified trees is a source of regret and sorrow for GunaiKurnai Traditional Owners. It also presents a lost opportunity to understand where the trees were originally distributed, their variability, and their associations with both landscape features and other kinds of cultural sites and materials. Despite their registration as cultural sites in some areas, there has been little research into culturally modified trees in Victoria, and none in Gippsland. As Jim Rhoads (1992: 202) has noted, 'most archaeologists do not find scarred trees to be intellectually engaging'. While dating of culturally modified trees has been carried

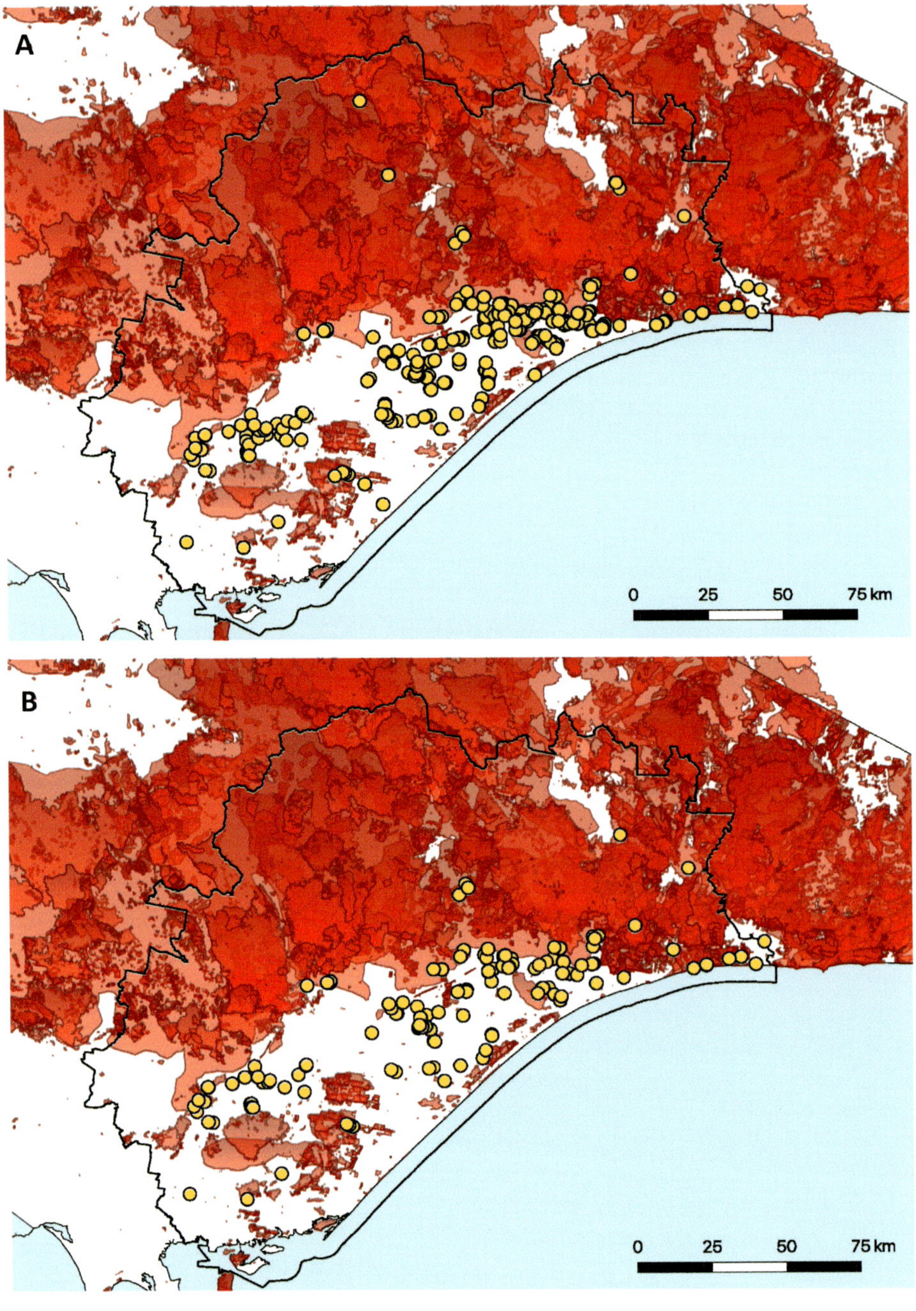

Figure 10.10. Distribution of Scarred Trees in the GKLaWAC RAP area and registered on the VAHR, relative to areas burnt by wildfires. The red areas represent the spatial extent of every documented wildfire since 1930. Each wildfire is mapped in the same shade of red in 50% transparency, so that the darker the red, the more fires have re-burned the same area over time. A: All registered trees since 1971. B: Registered trees that have not been inspected since their original recording. Their present state of preservation is not known. C: Registered trees that have been removed or destroyed. D: Registered trees that are still in their original locations (alive or dead, standing or fallen) (fire information and base map: from Data.Vic; Scarred Tree data: from VAHR).

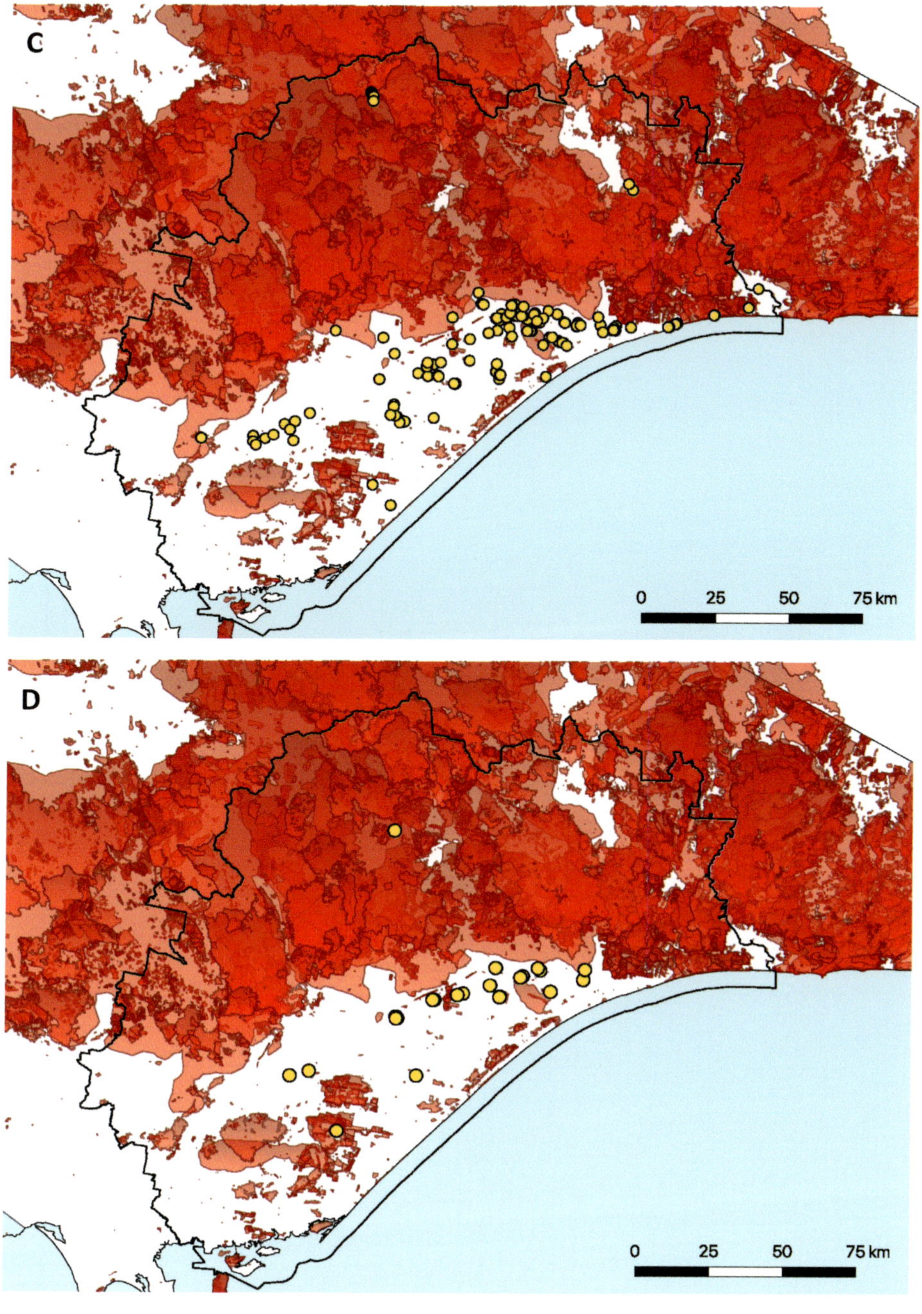
C
0
25
50
75 km
D
0
25
50
75 km

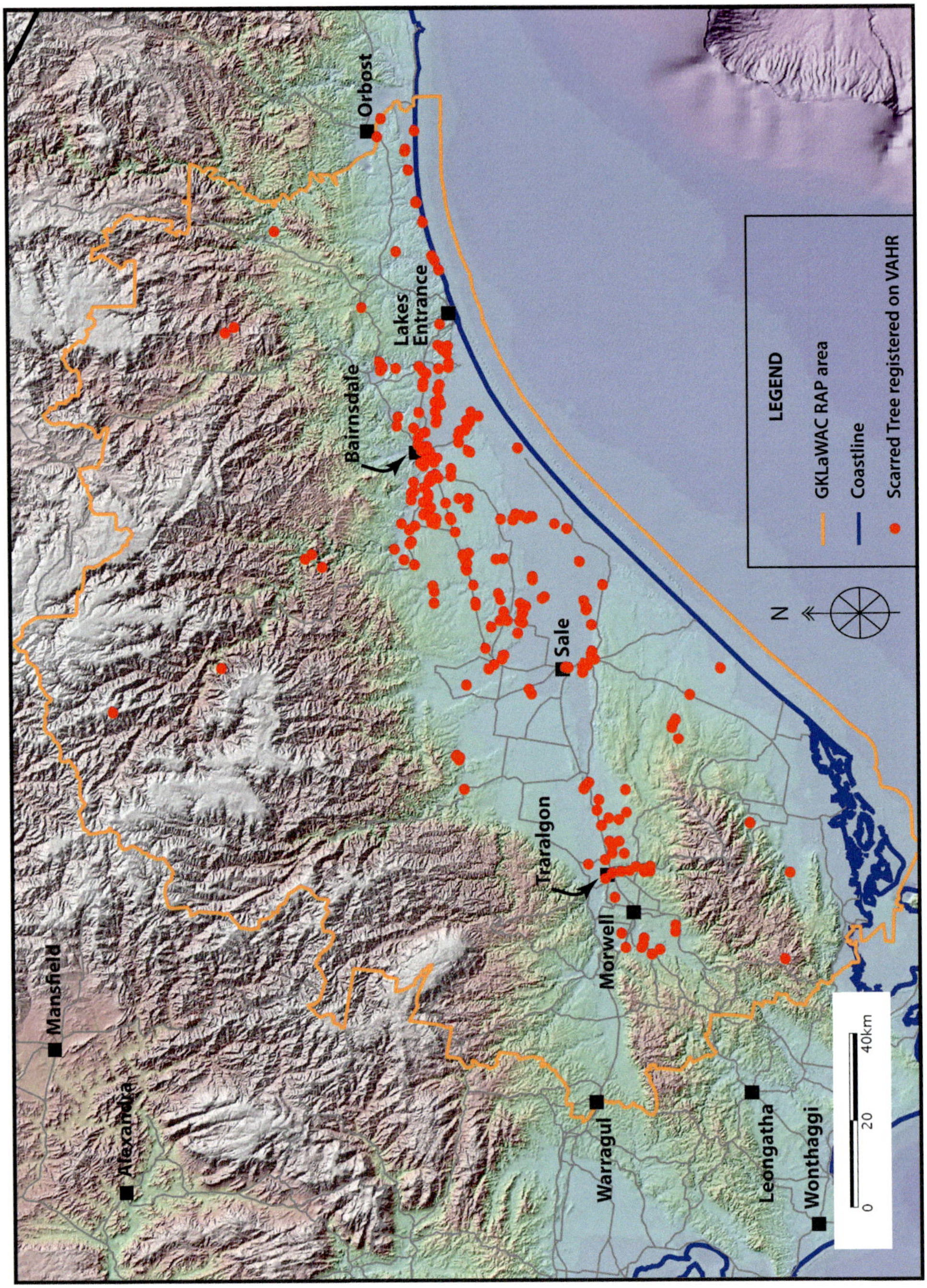

Figure 10.11. Scarred Trees in the GKLaWAC RAP area registered on the VAHR, showing their distribution relative to altitude. The lower, coastal and near-coastal plains are represented by the green hues (elevation data and base map: from GeoVic Digital Elevation 2004 layer; Scarred Tree data: from VAHR).

Figure 10.12. Culturally modified tree wrapped for protection during the 2015 Snowy River wildfires. A: The tree before wrapping and prior to the fire. B: The wrapped tree shortly after the fire (photos by Michael Douthat, courtesy of GKLaWAC).

out in the Northern Hemisphere, no such dating has been carried out in GunaiKurnai Country, and rarely has the antiquity of culturally modified trees ever been systematically attempted elsewhere in Australia (see examples in Ericsson *et al.* 2003; Johnson *et al.* 2022; Mobley and Eldridge 1992; Morrison and Shepard 2013; Navin Officer Heritage Consultants Pty Ltd 2015; Östlund *et al.* 2005, 2009).

A number of culturally modified trees, most dead and some fallen, feature in public spaces as a way of both preserving the trees, and of public display of local Aboriginal heritage. These trees from the Old Ancestors and those hidden in the forests of Gippsland have a story to tell. There is an urgent need to protect those that remain, to carry out further research, and to work with the GunaiKurnai to care for Country, to care for woodlands and forests, before further fires erase them from the landscape.

Chapter 11

Shells and Fire—Indicators and Effects

Katherine Szabó and Annette Oertle

INTRODUCTION

Writing an overview of the nature and impacts of fire on archaeological shell and shell deposits is more challenging than it may first appear, largely due to the fact that very little systematic research has been undertaken. The often casual way in which archaeologists may designate a particular shell or shell deposit 'burnt' may bely this, but in our experience, close questioning of the indicators used (e.g., colour changes, textural surface changes) reveals that the information relied upon is frequently incorrect, incomplete, or not backed up by experimental studies. There is no 'go-to' reference for discussion of evidence of burning on archaeological shell, and no paper or book which considers the various signatures and effects in a comprehensive or systematic way. What we attempt here, therefore, is to draw on the scattered, extant literature alongside our own experiments and observations to try and provide insights into the known and likely impacts of fire on archaeological shells and shell deposits.

Sporadic experimental studies conducted in various places around the world are attempting to understand the impact of fire on shell. Such studies are largely designed to help recognise methods of food processing (e.g., Aldeias *et al.* 2019; Milano *et al.* 2016, 2018), but some studies also aim to identify instances of the controlled use of fire to produce shell artefacts (e.g., Emory 1975: 20, 109 and Sillitoe 1988: 382 in Szabó 2005; Kozuch 2003). In some parts of the world, burnt shell is also used for the production of lime and mortar (Toffolo 2021), thus the deliberate burning of shell can happen at an industrial scale. Research to develop methods to distinguish between targeted burning (such as food processing) and incidental burning (such as exposed shell middens being impacted by larger-scale fires) is largely absent.

In Aboriginal and Torres Strait Islander archaeological sites, burnt shell is likely to occur as a result of one or other of three processes: (1) the cooking of shells for food preparation; (2) the incidental burning of shell midden through proximity to later campfires; and (3) the incidental burning of shell midden via landscape fires, whether natural wildfires or cultural burns. The difference between these three types of burning is primarily one of scale, with food preparation over a campfire having very local effects on a concentrated group of shells, and landscape fires impacting entire exposed archaeological shell deposits. These and other potential distinctions will be discussed further below.

From a materials science perspective, 'shell' is not a very useful category. Although all molluscan shells are produced in calcium carbonate ($CaCO_3$), there is a huge variety of different ways in which $CaCO_3$ crystals are enmeshed together among shell families. In the same way that a pine plank and a sheet of particle board might respond very differently to fire despite both being 'wood' and 'resin', a pipi and an oyster respond differently due to their dissimilar structures.

Given that, we start with a brief discussion of the major structural types of shell. We then summarise the current evidence for the impact of fire on these different structural types and issues associated with current knowledge. Finally, we look at some of the implications of the burning of shell for both archaeological analysis and heritage management.

STRUCTURAL VARIATION IN MOLLUSCAN SHELL

Mollusc shells are constructed in calcium carbonate ($CaCO_3$) with crystals of different sizes and shapes being laid down in various prescribed ways depending on the type of shell. Calcium carbonate comes in a range of forms, and in mollusc shells it tends to be one of two forms: aragonite or calcite. These two forms are associated with different types of crystal shapes, patterns and lattices, and these are referred to as microstructures. The best-known example of a shell microstructure is mother-of-pearl, also called nacre. In mother-of-pearl, tiny tablets of aragonite are arranged in stacks glued together by thin layers of organic proteins (Figure 11.1A). However, mother-of-pearl only occurs as an inner layer of shells and sandwiched to it as an outer layer is shell of a different microstructure—usually prismatic. Prismatic microstructures are comprised of long thin needles arranged side by side perpendicular to the outer surface of shell (Figure 11.1D). As with mother-of-pearl microstructures, individual crystals are glued together with organic proteins. A prismatic exterior and mother-of-pearl interior is characteristic of the warrener shell (*Lunella undulata*). Although mother-of-pearl also occurs in some bivalve (two-parted) shells such as marine and freshwater mussels, the structure of bivalve nacre is slightly different. Instead of stacks of crystal tablets being glued to each other ('columnar nacre'), in bivalves these tablets coalesce into very thin sheets which extend across the inner surface of the shell ('sheet nacre'). As is explored further below, the differences between these two structures means that they behave differently when impacted by fire.

In addition to nacreous/mother-of-pearl and prismatic microstructures, other common microstructures include (but are not limited to) crossed-lamellar and foliated. Crossed-lamellar structures are comprised of bundles of aragonite crystal tablets glued together in long rods. These bundles are interlocked together at angles to create a criss-cross or herringbone pattern (Figure 11.1B). This structure imparts strength to the shell and means that any fractures do not follow a straight line (as energy from force travels along crystal paths rather than breaking crystals) (Currey 1988). Crossed-lamellar microstructures are characteristic of a wide array of shells including pipis (*Donax deltoides*) and cone shells (Conidae). Fewer species of shell are comprised of foliate microstructures, but one notable type of shell in this category is the oyster (*Ostrea, Saccostrea* and *Crassostrea* species). Foliate structures are composed of thin, flat, pointed tablets made of calcite which are set together in an overlapping fashion like roof shingles (Figure 11.1C).

Before looking at the impacts of fire on the various microstructures and types of shell they are associated with, it is important to note that different microstructures contain different proportions of organic materials. These organic materials are comprised of proteins, glycoproteins, and polysaccharides which assist in building and maintaining the shell (Marin and Luquet 2004). The most organic-rich structures are nacreous/mother-of-pearl microstructures, where organic proteins encase each crystal as well as gluing together individual crystals and stacks (gastropods) or sheets (bivalves) of crystals. The microstructure

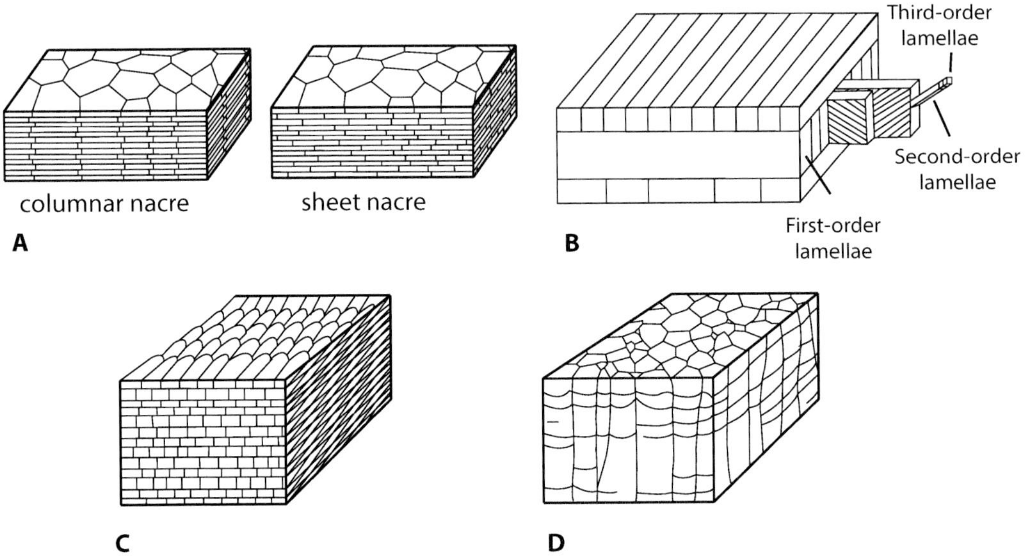

Figure 11.1. Schematic drawings of major microstructural types in mollusc shells: (A) columnar and sheet nacre. B: Crossed-lamellar. C: Foliate. D: prismatic (from Szabó 2017: 316).

with the lowest proportion of organic material is the crossed-lamellar type, where a significant amount of the strength of the structure is derived from the interlocked nature of the crystal bundles themselves.

RESPONSES OF SHELL TO BURNING

Burning is a complex combustion process whereby a number of physical and chemical transformations occur. Regarding shell, in the most general terms, calcium carbonate will disassociate at high temperatures to form carbon dioxide gas (CO_2) and calcium oxide (CaO). The CaO can then bond with water to form calcium hydroxide ($Ca(OH)_2$); a white, crystalline, highly-caustic powder commonly known as quicklime (Rye 1987 in Rude and Trinkle Jones 2012: 100). The reality, however, is more varied and complex as both fires and shells themselves are varied and complex. With regards to fire, temperature and duration will theoretically prompt different changes to shell at different rates, and the type of fuel will also change the nature of the fire. With regards to shell, the differing microstructural arrangements, amounts of organic proteins and basic mineralogy (i.e., aragonite versus calcite construction) will theoretically result in different types, rates and degrees of transformation. While we can assume levels of variability, experimental work to assess specifics has been sparse, patchy and sometimes flawed in its design or assumptions. Most notably, 'burning' and 'heating' are often conflated in research methodologies (e.g., Villigran 2014) but the very different chemical pathways means that furnace-heating versus experiments using fire should be considered entirely separately (e.g., Aldeias *et al.* 2019; Oertle 2019). Despite the complexity, recent high-quality work allows us to draw some meaningful conclusions and make some well-founded predictions about transformations.

All fresh shells contain water and organics to various degrees, and in contact with fire these are the first two parts of the shell to be affected (Gaffey *et al.* 1991). In burning experiments with pearl oysters (*Pinctada maxima*), where the meat had been harvested 3–5 years prior, Szabó and Koppel (unpublished data 2012) noted steam and the boiling of trapped liquid between nacre layers as the first visible sign of physical transformation. Measurements and weights of blood ark clams (*Tegillarca granosa*) showed that while the dimensions of whole shells increased as a result of delamination and the separation of layers of shell (see below), weight decreased through the loss of organics and water (Oertle 2019: 61). Currey (1979) has demonstrated that dehydrated shells are less strong than fresh shells of the same type. In archaeological terms this makes them more vulnerable to breakage and fragmentation.

The impact of the loss of the organic component within the shell depends largely on microstructural type (Glover and Kidwell 1993). In the case of prismatic structures, which are high in organics, exposure to fire sees surfaces lose integrity and sections of shell fracture away in blocks (Figure 11.2) (e.g., Oertle 2019: 96 for *Turbo setosus*). In the case of nacreous and foliate microstructures, where sheets and stacks of crystals are glued together by organics, the loss of these can result in the splitting of layers (Figure 11.3), thermal fracture and sometimes total disaggregation (e.g., Oertle 2019: 97 for *Turbo setosus*). Total disaggregation is reported for the experimental burning of mussel shells which were deliberately placed in the path of controlled backburning to investigate the impacts of wildfires on shell midden resources in North Dakota (Picha *et al.* 1991 in Haecker 2012: 139). The fire reached a maximum temperature of 399 °C and the mussel shells were noted to have disintegrated (Haecker 2012: 139).

Although crumbly and flaky mollusc shell is commonly encountered by those who analyse archaeological shell deposits, disintegration of this nature is not the only potential structural outcome of fire damage. In controlled experiments, Oertle (2019) identified structural transformations whereby individual crystals coalesced into homogenous blocky aggregates in Blood Ark Clam (*Tegillarca granosa*), oyster shell (*Crossostrea glomerata*—rock oyster) and the tropical Mud Mussel (*Geloina expansa*). This transformation is termed 'recrystallization', when the less stable aragonite form of calcium carbonate spontaneously transforms to its more stable form: calcite. In doing so the shapes and configurations of the crystals also change. This result was also backed up through chemical analysis (X-ray diffraction) which demonstrated the shift from aragonite to calcite construction (Oertle 2019). Although recrystallisation can be prompted by a variety of things, exposure to high temperatures is one of the main causes. Some texts cite very specific temperatures for recrystallisation for shell in general (e.g., 300–400 °C, see Faust 1950; Passe-Courtin *et al.* 1995), but Oertle's (2019) results from experiments across a range of species showed that both recrystallisation temperatures and rates were quite varied. For example, Mud Mussel (*Geloina expansa*) nearly completely recrystallised after only five minutes on hot coals, without exposure to flames, but Gold-Lip Pearl Oyster (*Pinctada maxima*) still retained some original aragonite mineralogy after ten minutes in direct flames measuring over 600 °C (Oertle 2019: 73–74, 92–94).

A further potential change, mentioned briefly above, is the complete transformation of shell material with the physical remnant being powdery, bright white calcium oxide. Oertle (2019: 96) noted this for *Turbo setosus* (a close tropical relative of warrener shell) after only

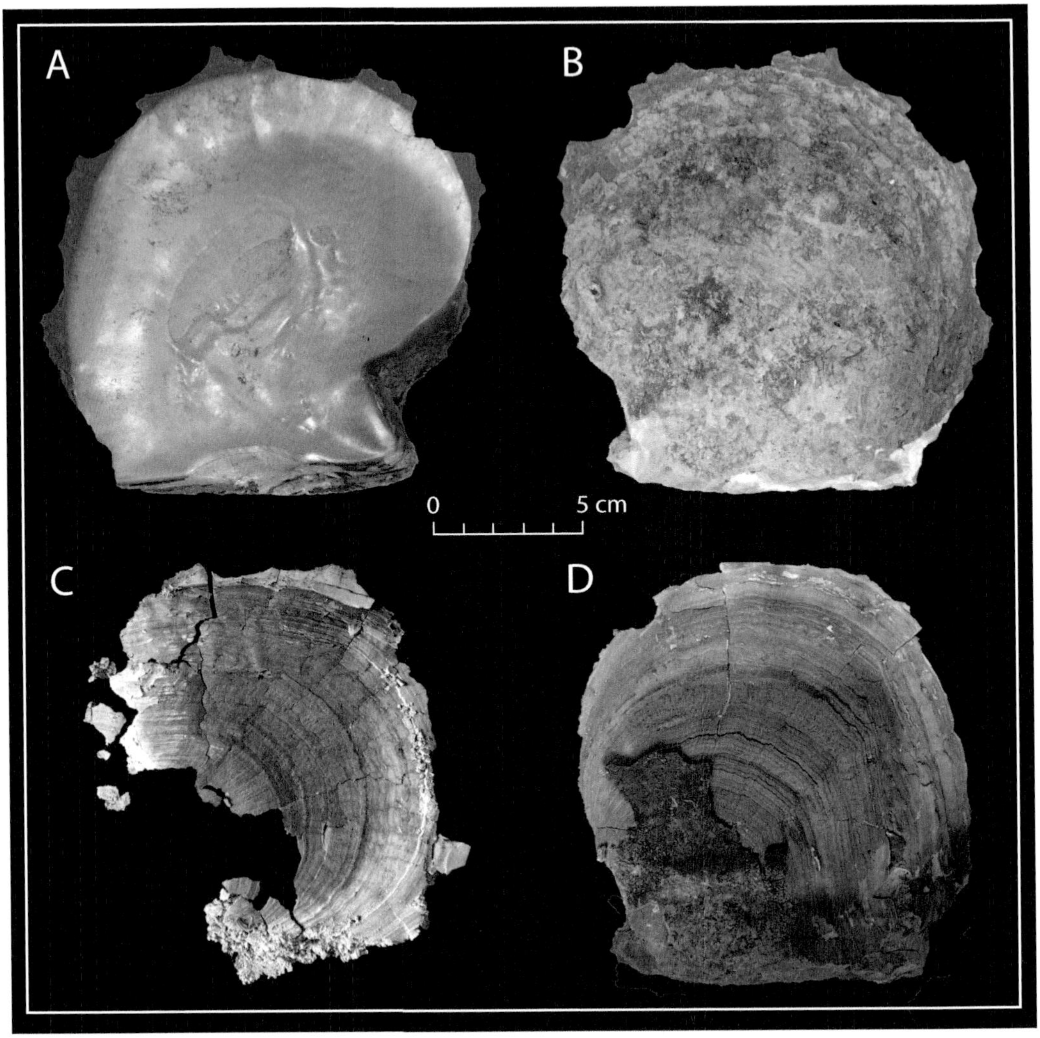

Figure 11.2. Results of controlled burning experiment of pearl oyster (*Pinctada maxima*) showing blocky fracture and disintegration of the outer prismatic layer of shell (from Oertle 2019: 361).

ten minutes in contact with the flames of a fire. Interestingly, samples of *Turbo setosus* heated to high temperatures (600 °C) in an oven (as opposed to an open fire) did not go through the same physical or chemical transformations and no calcium oxide was observed (Oertle 2019: 99).

Although the common pathway acknowledged across the relevant literature is for aragonite to first transform to calcite, and then to calcium oxide in the presence of high temperatures, there is another potential transformation. The aragonite created by molluscs when building their shells is termed 'biogenic aragonite' as it is biologically created. Naturally occurring aragonite in geological contexts is termed 'geologic aragonite', but there is a further type: 'pyrogenic aragonite'. This type of aragonite is formed through burning and shows visually

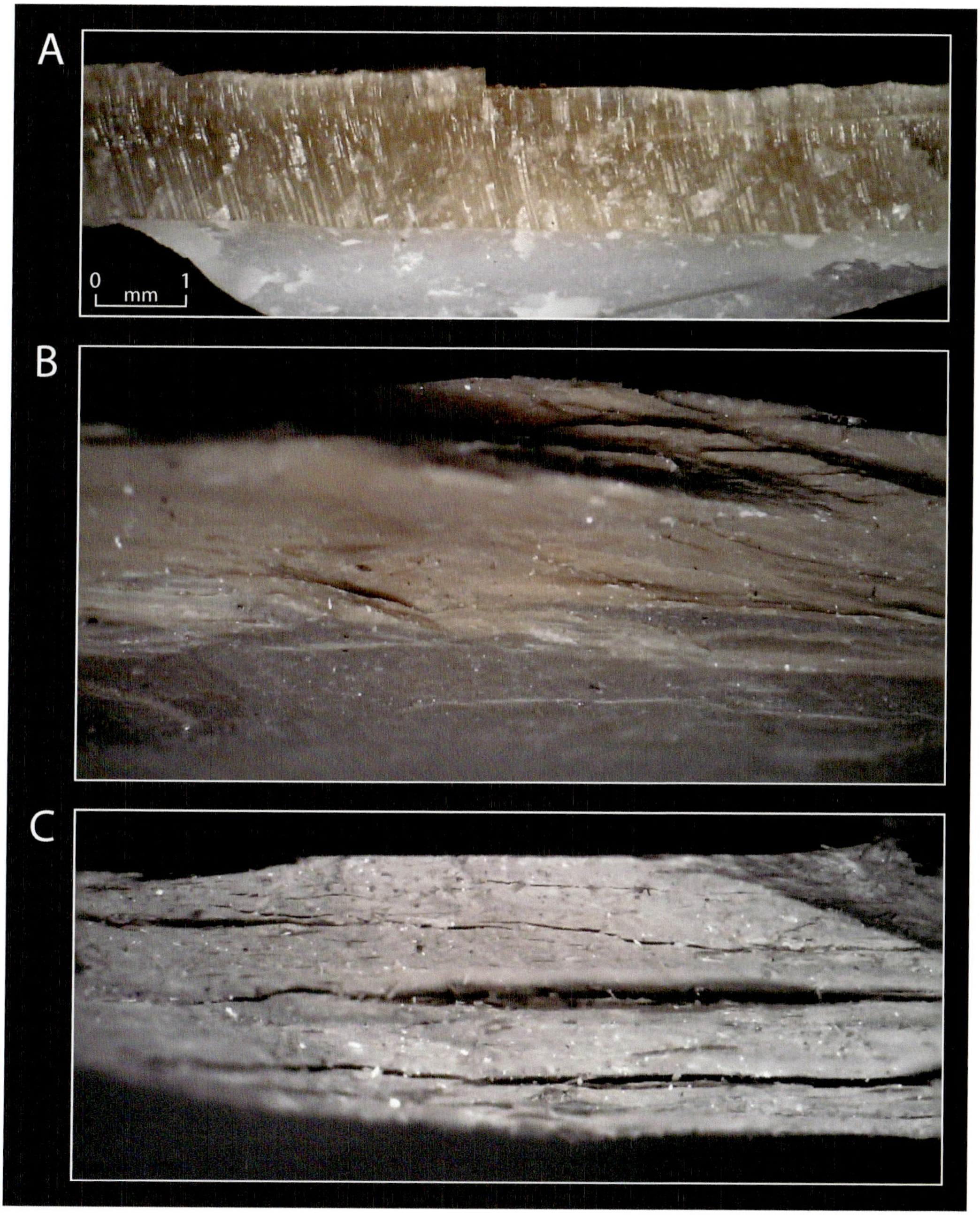

Figure 11.3. Results of controlled burning experiments with Gold-Lip Pearl Oyster (*Pinctada maxima*). The left image shows a cross-section of both prismatic (top) and nacreous (bottom) layers largely intact. The images on the right show the splitting apart of the nacre layer in sheets after burning. Scale in mm intervals (photographs: Annette Oertle).

under high magnification as needle-shaped ('acicular') crystals, often present in clusters (Toffolo 2021: 155). Pyrogenic aragonite does not form under all high-temperature burning conditions of $CaCO_3$ and its mechanics are not fully understood (Toffolo 2021: 155). In burning experiments conducted by Oertle (2019), pyrogenic aragonite needles were noted as having formed in some Blood Ark Clam (*Tegillarca granosa*) samples, although interestingly only when indirect burning (resting on hot coals) was used to transform samples. Combined information from Toffolo (2021) and Oertle (2019:64) suggests there are variables beyond temperature and duration at play in the formation of pyrogenic aragonite and very specific conditions are required.

Samples can be put under a microscope or tested to assess their chemical composition, but most archaeologists will identify shell as 'burnt' based on macro changes in colour. In some instances, the shell will be noted to be grey, in others white, and in others black, however systematic evaluation from experiments shows that different temperatures, fire types, duration of exposure and shell types will yield very different results (Aldeias *et al.* 2019; Oertle 2019). A further complication is that various other processes of transformation and decay can lead to surfaces which look similar. A powdery, white surface on a shell could also indicate weathering and acid dissolution, where exposure to rainwater and the elements prompts the rapid deterioration of organic 'glue' components and dissolves the calcium carbonate crystals. Grey or blackened shell can also occur under natural decay conditions given the right chemical and microbial circumstances (Callender *et al.* 1994). Thus, discolouration alone is not a reliable marker for burning.

ISSUES ASSOCIATED WITH BURNT SHELL

Although the impacts of burning on shell are far from perfectly understood, and shell responses vary dramatically due to the different nature of shell structures as well as fires, there are some general impacts which can be noted. All of these are either likely to happen if archaeological shell deposits are impacted by wildfires or must be tested for before further analysis continues.

The most obvious impact of a large-scale fire on archaeological shell deposits is physical damage, and possible complete destruction, of the shells themselves. As discussed above, species which have structures that are richer in organics or are otherwise more vulnerable are likely to delaminate, fracture or completely disaggregate. As mentioned earlier, the complete disaggregation of mussels was noted in control samples placed in front of controlled back-burning fires in North Dakota (Picha *et al.* 1991 in Haecker 2012: 139). Dehydration, the reduction of the organic fraction and recrystallisation will all make shells more vulnerable to fracture and post-fire weathering leaving remaining shell cultural resources susceptible to weather and other impacts. Relevant to GunaiKurnai Country and other parts of Victoria is the fact that freeze-dry cycles on remains following burning will also intensify damage (see Villagran and Poch 2014 data for Argentina). Given the variability between different types of shell we can also expect that any remaining sample will be biased, with more robust types of shell outlasting more fragile counterparts.

While all of this sounds very logical, and is backed up by burning experiments conducted in controlled conditions (e.g., Milano *et al.* 2016, 2018; Oertle 2019), there are circumstances when the unexpected occur. In deposits dating from 35,000–50,000 years ago at the Niah Caves in Borneo, the best-preserved shells were burnt black and recrystallised (Reynolds *et al.* 2013: 166), while the rest of the shell sample was powdery, disaggregated and was noted to crumble when touched during excavation (Harrisson 1957: 133 in Reynolds *et al.* 2013: 165). It is possible that the protection of dry, cave sediments afforded some physical protection whilst the removal of the organic fraction slowed normal decay processes, but further investigation and experimentation would be required to test this.

Experimental and analytical work to date has demonstrated the dramatic transformations in colour and structure when shells are directly impacted by hot fires. When shells are away from direct fire, either at the margins of a fire or buried beneath a burnt landscape, the impacts are less clear. In these circumstances there can be little to no visual evidence of burning (Koppel *et al.* 2017; Milano *et al.* 2016, 2018) and no or incomplete recrystallisation. Such shells which have been only lightly impacted or partially transformed may be unwittingly included in analyses and produce spurious or perplexing radiocarbon dates or other stable isotope analysis results (Larsen 2015). Given this, the testing for fire/heat-induced transformations is critical prior to undertaking further analysis and is routinely done by some radiocarbon laboratories (e.g., Waikato Radiocarbon Dating Laboratory 2017).

The standard approach to testing for recrystallisation is generally X-ray diffraction (XRD), which can determine the occurrence and proportions of aragonite, calcite, and other minerals if desired. If, however, a sample has been impacted by heat or fire, but not to the point of recrystallisation, XRD is of little help. A different technique is showing promise in these circumstances: amino acid racemisation (AAR). All living things contain amino acids and after death these acids change orientation. The rate of change is different for various types of amino acids and is also variously impacted by environmental context, particularly temperature. Recent research on shells has determined that, in the presence of the sharply-elevated temperatures of a fire, two of the most common and routinely-measured amino acids will behave very differently: aspartic acid will racemise much more quickly than glutamic acid (Brooks *et al.* 1991; Crisp 2013; DeMarchi *et al.* 2011; Koppel *et al.* 2017). Thus, even in circumstances where there has been no visual evidence of burning, fire-affected shell can be isolated and excluded from further analysis (Koppel *et al.* 2017). This technique may be particularly useful in circumstances where sub-surface remains are potentially impacted by landscape and other fires (e.g., Bennett 1999 for buried bone).

The processes and exchanges involved in the transformation from aragonite to calcite as a product of burning will not just complicate radiocarbon dating (^{14}C) but also the testing of different atomic weights of carbon and other elements such as nitrogen and oxygen. Elements with variations having different atomic weights are called isotopes. Various isotopes and their ratios can inform on many different things such as the nature of past environments and diets, and thus if a sample is burnt and fire-altered, results may be compromised (e.g., Robinson and Kingston 2020 for tooth enamel). The extent of the impact of this on shell samples is not well understood, but controlled experiments with fish otoliths (also made of aragonite) have demonstrated the extent to which fire can impact upon accurate results (Andrus and Crowe 2002).

A further complication with regards to radiocarbon dating (and possibly other forms of stable isotope analysis) is highlighted by Lindauer *et al.* (2018) based on the experimPlease change to Katherineal research of Zazzo *et al.* (2012). They point out that if 'old wood' is used as a fuel for roasting shellfish, then the shells can uptake some of the chemical signatures of the wood. This can produce radiocarbon dates which are artificially old (Lindauer *et al.* 2018: 530). Despite these unexpected results, the experiments conducted in Lindauer *et al.* (2018) suggested that burning did not affect the reliability of radiocarbon dates in two common species of Indo-Pacific shell. Clearly more work remains to be done.

BURNT SHELL AT A LANDSCAPE LEVEL: IMPLICATIONS FOR ARCHAEOLOGY AND HERITAGE MANAGEMENT

All of the foregoing information demonstrates that fire has a significant impact on molluscan shell, but also that impacts are highly variable and not always well understood. Fire will cause colour changes, although these vary depending on both the nature of the fire and the shell, and shells will also lose structural integrity in various ways. Mineralogy will often transform, and this in turn may impact upon the ability of particular techniques, such as radiocarbon dating, to produce robust results.

The majority of research into the impact and signatures of fire on archaeological shell to date has been developed from a perspective of investigating food processing techniques. As such, highly localised fire signatures form the basis of experiments and often the identification of burnt faunal (e.g., bone or shell) and/or stone artefacts are taken as proxy evidence of human agency. This logical shorthand might work in many parts of the world, but in places like parts of the United States and Australia where natural fires are common, the association can be assumed to be less strong. To complicate matters further, in between large, natural landscape fires and small localised domestic fires, are those cultural practices which use fire to manage aspects of the landscape. In this category sits the Aboriginal uses of fire to manage and maintain landscapes; and elsewhere landscape-scale burning can also be found, such as in areas where swidden (slash and burn) farming formed the basis of traditional agriculture (e.g., Padoch *et al.* 1998 for Borneo) and the increasingly strong evidence for pre-Columbian controlled landscape burning in North America (e.g., Abrams and Nowacki 2008; Walsh *et al.* 2018). As with any complex archaeological question, investigation of the agent, extent, nature and impact of fire can only be successfully tackled using multiple lines of evidence.

With specific regards to archaeological shell, knowledge accumulated to this point suggests the following impacts and points of caution in relation to large-scale fires:

1. Evidence of burning and ensuing fragmentation of shells may relate to food processing, larger-scale burning or both. The most likely way of distinguishing between the two is to consider the extent of burning spatially and assess this alongside other lines of evidence.
2. Different types of shell will react differently to fire and so any remaining shells must be expected to be a biased sample: the most robust and organic-poor shells are expected to out-survive organic-rich species of more vulnerable construction.
3. Shell that has been burnt is unlikely to be a reliable material for radiocarbon dating or for stable isotope analysis. Samples should therefore be selected with caution.
4. In some scenarios, such as coastal archaeological shell deposits sitting atop natural shell-bearing structures (e.g., beach ridges), it may be virtually impossible to separate

naturally occurring versus archaeological shell after a large-scale fire. Speiss (1988) points to an analogous problem with charcoal. The normal indicators used by archaeologists, such as types of shell breakage for meat extraction and the selection of particular types of shell, will be muddled and biased after severe fire damage, thereby creating uncertainty.

5. Post-fire conditions tend to be characterised by high levels of erosion (Ryan, Koerner *et al.* 2012: 12), and heightened sediment movement around already fragile shells will exacerbate deterioration. Shell surfaces will be more likely vulnerable to abrasion, whole shells and fragments less resistant to crushing and compaction, and edges will be frailer. Damaged and porous surfaces will also create a greater surface area for acid dissolution prompted by rainwater, also accelerating decay.

It is presently unclear to what extent buried shell may be impacted by landscape fires. Intense heat above buried shell may still prompt physical and chemical changes like recrystallisation. Despite this, if they are shielded from direct flames and post-fire exposure, erosion and weathering the chances of survival seem much better. Likewise, archaeological shell deposits shielded by natural structures such as caves and rock shelters may be afforded protection during a landscape fire. Although this preserves aspects of material heritage, it also presumably biases the archaeological record across the landscape with more vulnerable archaeological material disappearing from analytical view.

Based on laboratory-based and controlled experiments, it is possible to hypothesize about the impacts of landscape fires on archaeological shell resources, but the very control inherent in experimental studies sometimes makes their inferences limited. Landscape processes are complex with many factors combining in sometimes mercurial ways. Adding Indigenous and Western fire and landscape management on top of this introduces a myriad of further variables. Given this, it would seem valuable to test understandings on the ground by assessing the survival of known archaeological shell resources after landscape burns, and also to deliberately place experimental shell and other materials of interest in the path of Indigenous and fire service deliberate burns (i.e. cleaning Country and prescribed burning) to systematically assess impacts.

PART 3
Understanding the Impact of Fires on GunaiKurnai Cultural Heritage Sites: Past, Present and Future

Previous page: Wildfire front, 2019–2020 Black Summer fires, burning near coast just east of Bairnsdale, East Gippsland. Photograph by Joanna Fresløv.

Chapter 12

Landscape Fires and Cultural Sites in GunaiKurnai Country

Jessie Buettel, Stefania Ondei, Bruno David, Joanna Fresløv and Russell Mullett

Fire is a feature of Australian landscapes, but while many Australian vegetation communities have adapted to and evolved with its presence (e.g., *Eucalyptus* trees requiring fire to regenerate), the recent increase in the extent and severity of wildfires (see below), especially in fire-sensitive landscapes, has left devastation in its wake. With this increase in fire occurrence and area burned comes an increased risk of damage to irreplaceable cultural sites.

It is therefore important to consider how and why the impact of fire might be different across the landscape. Because cultural sites are distributed unequally across GunaiKurnai Country, it is likely that the impacts of fire will not be the same everywhere. Rather, those impacts will be dependent on the local-scale characteristics of the sites, such as vegetation type, land-use type, and the composition and structure of cultural sites and their artefacts.

While our focus is the GKLaWAC RAP area, we sometimes address a slightly broader geographical area in this chapter, from the east coast to the south coast, central Victoria and the Australian Alps, to better take into account the nature and effects of wildfires across the land systems within and immediately surrounding the GKLaWAC RAP area.

Unsurprisingly, and the land-use biases noted above notwithstanding, the number of registered cultural sites burnt corresponds strongly with the amount of area burnt through time, with the larger fires generally indicating a higher number of burnt cultural sites. Since 1930, a total of 970 cultural sites have been burnt at least once, irrespective of fire type (Table 12.1). Over 32% (n = 870) of all registered cultural sites (n = 2698) have been burnt by wildfires at least once, with 11% (n = 298) burnt once by prescribed burning (Table 12.1). In this context, it is worth noting that during pre-planning for prescribed burns, which usually cover relatively small areas, care has generally been taken to avoid burning areas immediately around cultural sites. The potential impacts on cultural heritage are usually checked well ahead of prescribed burns. However, such checking has mainly focused on sites along roads and tracks to ensure any track-clearing will not impact on sites, rather than a consideration of the impacts of the burns themselves on sites away from the tracks.

When looking at how the number of registered cultural sites burnt by wildfires changes through time, the number decreases from 743 now-registered sites burnt between 1930–1999 (i.e., at an average rate of 11 registered sites/year), to 472 between 2000–2020 (22 registered sites/year), matching the greater area of land burnt in the 20th century but at triple the average annual rate of burning in the 21st (see Chapter 5) (Figure 12.1).

Table 12.1. Number of registered cultural sites burnt by wildfires and prescribed burns since 1930 in GunaiKurnai Country. Note that each site is represented only once in this table, though we know that some sites were burnt by either fire type more than once. The total number of registered cultural sites burnt (far right column) is not the sum of the number burnt by wildfire plus prescribed burning, as the same site could have been burnt by both fire types over time.

Site Type	Total number of sites	Number burnt by wildfire	Number burnt by prescribed burning	Total number of sites burnt
Aboriginal Ancestral Remains (Burial)	10	1	0	1
Aboriginal Ancestral Remains (Burial) and Aboriginal Historical Place	1	0	0	0
Aboriginal Ancestral Remains (Burial) and Artefact Scatter	6	0	0	0
Aboriginal Ancestral Remains (Burial) and Artefact Scatter and Shell Midden	4	0	2	2
Aboriginal Ancestral Remains (Burial) and Shell Midden	1	0	0	0
Aboriginal Ancestral Remains (Reinterment)	1	0	0	0
Aboriginal Cultural Place	1	0	0	0
Aboriginal Historical Place	13	5	1	5
Aboriginal Historical Place and Artefact Scatter	1	0	0	0
Artefact Scatter	1645	712	207	756
Artefact Scatter and Earth Oven	1	1	0	1
Artefact Scatter and Grinding Grooves	2	0	0	0
Artefact Scatter and Hearth	4	0	0	0
Artefact Scatter and Hearth and Shell Midden	1	0	0	0
Artefact Scatter and Hearth and Subsurface	1	0	0	0
Artefact Scatter and Quarry	3	3	1	3
Artefact Scatter and Rock Art	1	1	0	1
Artefact Scatter and Rock Art and Subsurface	2	2	1	2
Artefact Scatter and Shell Midden	145	4	25	28
Artefact Scatter and Shell Midden and Subsurface	2	0	0	0
Artefact Scatter and Subsurface	37	10	0	10
Bora Rings	1	0	0	0
Grinding Grooves	20	0	0	0
Grinding Grooves and Quarry	1	0	0	0
Low Density Artefact Distribution	163	39	19	47
Mound	1	1	0	1
Object Collection	34	2	4	4
Quarry	10	10	1	10
Scarred Tree	387	74	24	81
Shell Midden	195	4	13	17
Shell Midden and Subsurface	4	1	0	1
Total sites	**2698**	**870**	**298**	**970**

In contrast, the number of registered cultural sites burnt by prescribed burning is remarkably similar over the two periods (1930–1999: 70 years, n = 186 cultural sites; 2000–2020: 21 years, n = 165 cultural sites), despite the greater extent of area burnt by prescribed fires between 1930–1999 (6721km^2 versus 3638km^2) (Figures 5.1, 6.1). However, even though slightly more registered cultural sites were burnt by prescribed burns between 1930–1999, on average three registered cultural sites were burnt per year (minimum = 0; maximum = 33) during those 70 years, compared with an average of eight burnt per year in the 21 years between 2000–2020 (minimum = 0; maximum = 456). In other words, in the past 21 years sites have been burnt through prescribed burning three-times as fast as in the previous 70 years. In the past two decades, the years that had particularly high numbers of registered cultural sites burnt by prescribed burning include: 2004 (33 sites), 2005 (27 sites), 2012 (25 sites), 2013 (19 sites), and 2019 (13 sites).

Each year since 1980 has seen at least one registered cultural site burnt by prescribed fires, with 1986 and 2020 the only exceptions in that 41-year period. The number of registered cultural sites burnt by prescribed burning was consistently close to 100 per decade across the past four decades, despite the occurrence of two major wildfires in 2007 and 2019–2020 and the large number of cultural sites burned as a result, compared to the prior two decades (1980s–1990s; Figure 12.2). This could be because prescribed burns were largely undertaken in the same locations each decade, as shown in Figure 5.3.

However, the greater number of registered cultural sites burnt by wildfires between 1930–1999 is largely a result of the two large-scale wildfire events of the 1930s and 1960s (Figure 12.2). These wildfires were massive in extent, in particular the Black Friday fires of 1939 that remain unmatched in extent and severity across the fire history record to date (see Chapter 5, Figures 5.1, 5.2). Indeed, the years that experienced these large-extent wildfire events contribute significantly to the numbers of registered cultural sites that burned by wildfire in each decade (Figure 12.2).

The four largest wildfire events each burnt on average 237 registered cultural sites (minimum = 125, maximum = 456; Table 12.2). By some margin, the Black Friday fires were the most extensive, burning 456 registered cultural sites, followed by the Great Divide fires of 2007, which burnt 211 registered sites, followed by the Gippsland fires of 2019–2020 with 165 registered sites, and the Gippsland fires of 1965 with 154 registered sites (Table 12.2). Surface Artefact Scatters have been burnt in high numbers during each of the large wildfire events, primarily because they are the most numerous and widespread of the cultural site types (Table 12.2).

Some wildfires and prescribed burning events can occur in the same location(s) year to year. This means that some cultural sites may be burnt more than once in a given year, decade, or century. The more times a cultural site is burnt, the greater the likelihood of incremental damage or loss of both artefacts and sites. Indeed, of the 870 registered cultural sites that were burnt by wildfires since 1930 (Table 12.1), 56% (n = 490) of these were burnt once, 21% (n = 182) twice, 18% (n = 157) three times, 4% (n = 34) four times and 1% (n = 7) five times. Of the seven registered cultural sites that were reported to have been burnt by wildfires five times, since 1930, six are of the site type Artefact Scatter, and one is an Artefact scatter and Quarry (Table 12.3). One registered cultural site (Artefact Scatter) was burnt by prescribed burning five times.

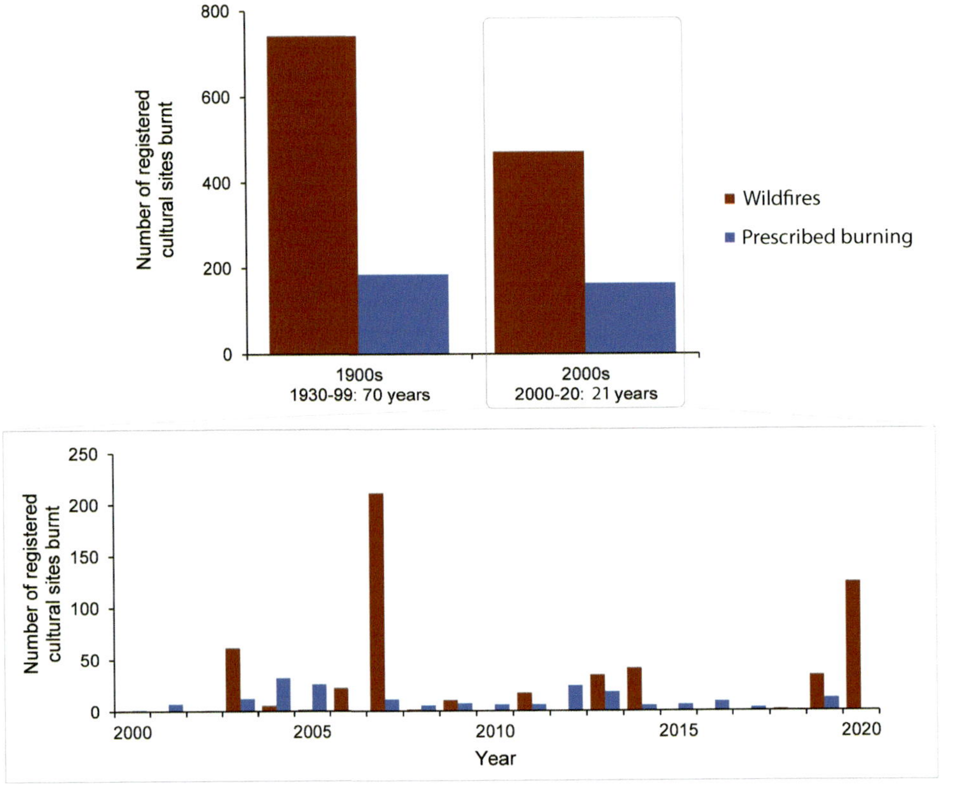

Figure 12.1. Number of registered cultural sites burnt by wildfires (brown bars) and prescribed burns (blue bars) between 1930–1999 vs 2000–2020. The lower graph shows the number of registered cultural sites burnt each year by wildfires and prescribed burns since 2000. Note that registered cultural sites are only counted once, even if they have been burnt by either fire type multiple times within each time period.

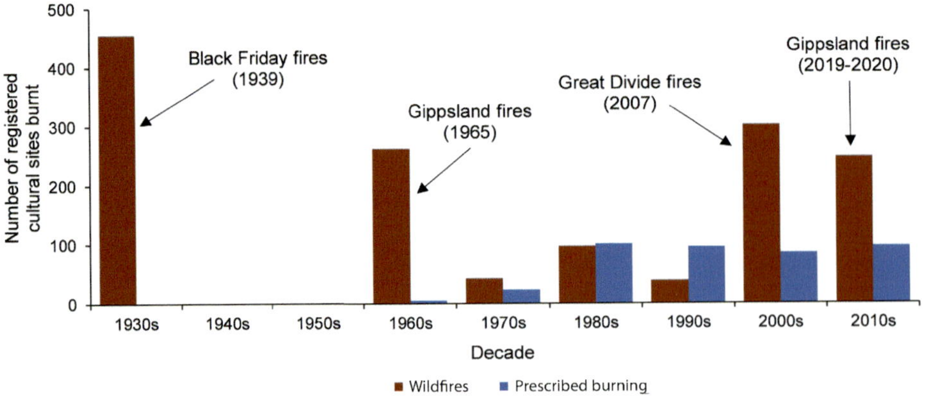

Figure 12.2. Number of registered cultural sites burnt by wildfires (brown bars) and prescribed burns (blue bars) across each decade in GunaiKurnai Country. The four largest fire events are indicated by arrows to their corresponding decades, with the year they occurred in brackets.

Table 12.2. Number of registered cultural sites that were burned by the four major wildfire events: Black Friday fires (1939), Gippsland fires (1965), Great Divide fires (2007) and Gippsland fires (2019–2020).

Site type	Black Friday fires (1939)	Gippsland fires (1965)	Great Divide fires (2007)	Gippsland fires (2019–2020)
Aboriginal Ancestral Remains (Burial)		1		
Aboriginal Historical Place	1		2	1
Artefact Scatter	416	106	194	136
Artefact Scatter and Earth Oven	1			
Artefact Scatter and Quarry	2		1	1
Artefact Scatter and Rock Art		1		
Artefact Scatter and Subsurface	5	2		2
Artefact Scatter and Subsurface and Rock Art				2
Low Density Artefact Distribution	10	15	3	11
Mound	1		1	
Object Collection				1
Quarry	6		1	6
Scarred Tree	14	29	9	4
Shell Midden				1
Total	**456**	**154**	**211**	**165**

Across all registered cultural sites, Artefact Scatters are the site type most often burnt more than once by wildfires, with 371 sites burnt once, and a total of 380 burnt twice or more (Table 12.3). They are also the most represented site type burnt more than once by prescribed burning, with 117 registered Artefact Scatters burnt once, and 106 burnt twice or more. Scarred trees are the next most frequent cultural site burnt by both wildfires and prescribed burning events (see Chapter 10), followed by Low Density Artefact Distribution and Artefact Scatter and Shell Midden (Table 12.3). Of interest, Artefact Scatter and Shell Midden, and Shell Midden, sites were burnt more often by prescribed burning than wildfires, probably due to their high numbers along the coast including in and near highly frequented public parks and reserves where fire management is heightened. There were also two Aboriginal Ancestral Remains (Burial) and Artefact Scatter and Shell Midden sites burnt once by prescribed burning but not wildfires (Table 12.3).

Table 12.3. Numbers of registered cultural sites that were burned by wildfires and prescribed burning, and the number of times they were burnt. Each site is represented only once within each fire type, with some sites burnt by wildfires also possibly having been burnt by prescribed burning.

Site type	Number of times burnt									
	Wildfires					Prescribed burning				
	1	2	3	4	5	1	2	3	4	5
Aboriginal Ancestral Remains (Burial)	1									
Aboriginal Ancestral Remains (Burial) and Artefact Scatter and Shell Midden						2				
Aboriginal Historical Place	3	2				1				
Artefact Scatter	371	148	154	33	6	117	53	29	7	1
Artefact Scatter and Earth Oven	1									
Artefact Scatter and Quarry		1	1	1	1			1		
Artefact Scatter and Rock Art	1									
Artefact Scatter and Rock Art and Subsurface	2					1				
Artefact Scatter and Shell Midden	4					24	1			
Artefact Scatter and Subsurface	6	4								
Low Density Artefact Distribution	31	8				14	4		1	
Mound		1								
Object Collection	1		1			2		2		
Quarry	5	4				1				
Scarred Tree	59	14	1			17	6	1		
Shell Midden	4					13				
Shell Midden and Subsurface	1									
Total	**490**	**182**	**157**	**34**	**7**	**192**	**64**	**33**	**8**	**1**

CONCLUSION

It is clear from the results presented in this chapter that more land area, and consequently more cultural sites, have been burnt since the 1980s than previously. This is true for both wildfires and prescribed burns. This accelerating rate of fire-damage to cultural sites shows no signs of abating, if these statistics are any indication. On the contrary, the period since the beginning of the 2000s has seen a further acceleration of the incidence of wildfires across the GKLaWAC RAP area. With this intensification of fires across the landscape, increasing numbers of cultural sites have been burnt and reburnt. We will return to these issues in Chapter 15.

Chapter 13

Archaeological Surveys in GunaiKurnai Country

Robert Skelly, Bruno David, Joanna Fresløv and Russell Mullett

A number of archaeological surveys have been undertaken following major wildfires in GunaiKurnai Country. These surveys have aimed to document the characteristics and distribution of archaeological sites in the fire-affected areas, and to make recommendations for their improved protection from the impacts of fires. Following the Caledonia Area wildfire in the Southern Alpine Region from December 1997–January 1998, David *et al.* (1998) assessed the impacts of heat from the fires and fire-suppression methods on Aboriginal sites. A sample of the burnt area was surveyed, with a focus on waterways, ridgelines and saddles which were assessed as likely places for Aboriginal land use and settlement. David *et al.* (1998) found that the fire improved the effectiveness of site identification without biasing an assessment of site distribution. The study found that radiant heat in high-intensity burn areas had detrimental impacts on surface stone artefacts, with one-third of the stone artefacts identified showing some form of fire damage (see Chapter 7 for a discussion of indicators).

Following wildfires in the Victorian southeastern alpine region in 2002–2003, Aboriginal Affairs Victoria (AAV) conducted a targeted investigation of the effects of the fires in what were deemed to be culturally sensitive areas, in areas where fire-suppression works had occurred, and in locations of proposed rehabilitation works resulting from the fires. These field surveys took place mainly in the High Country at Dinner Plain, Horsehair Plain, Mount Hotham, Mount McKay, Bogong High Plains, Omeo, Orbost, Gelantipy, Suggan Buggan and Limestone Creek (Aboriginal Affairs Victoria 2003: 2, 6, 10). The surveys that took place in areas where mechanical fire-suppression activities had taken place (see below) found 21 Aboriginal sites in the disturbed areas (Aboriginal Affairs Victoria 2003: 5). Mechanical fire-suppression, especially from the movement of heavy machinery along ridge-lines and clearing of ground-cover, was thus found to have damaged many Aboriginal sites, with some being completely destroyed. A key finding from the report was that fire-control lines are likely to be positioned on culturally sensitive landforms (i.e., the ridge-lines which were ancient travel routes), and, as a consequence, that fire-control lines have had disproportionately high impacts on the integrity of Aboriginal sites, often critically damaging or totally destroying sites (Aboriginal Affairs Victoria 2003: 5). Ten further sites were found in areas where vegetation had been cleared for rehabilitation works, such as areas with 'dog fencing' (being fence-lines erected to keep wild dogs from attacking stock). Another 30 sites were also found in undisturbed areas thought likely to contain sites and with good ground visibility following the fires (Aboriginal Affairs Victoria 2003: 9–10).

At the same time as the AAV surveys were being undertaken, Parks Victoria (PV) and the Victorian government's Department of Sustainability and Environment (DSE) commissioned a post-wildfire cultural heritage investigation of the 2002–2003 Victorian northeastern alpine fire region, mostly outside but incorporating small sections of GunaiKurnai Country (Kelly

2003). As with the AAV surveys, they focussed on areas where fire suppression activities in the High Country and foothills at Mounts Beauty and Buffalo had disturbed the land, as well as on areas in hill country around the townships of Buckland, Corryong, Eldorado and Stanley (Kelly 2003: 5–6). The survey sampled landforms that were thought to contain archaeological sites, such as lower river valleys and peripheral slopes; middle altitude watercourses and valleys, upper ridges, spurs and slopes; and High Country valleys, ridges, spurs and peaks (Kelly 2003: 6). Twelve Aboriginal sites were identified during the surveys (Kelly 2004: 7). The study also included some experimental research on the effects of fire on quartz (the main raw material used for artefacts in the northeast) (Kelly 2003: 7). It concluded that heat from low-intensity burns had minimal impacts (discolouration) on stone artefacts, whereas high-intensity burns increased their brittleness and left them susceptible to secondary fracturing (Kelly 2003: 7). Where Aboriginal sites were located in areas impacted by mechanical fire-suppression, damage to sites was severe (Kelly 2003: 8).

Following the 2002–2003 fires, Victorian government agencies established the Public Land Ecological and Cultural Bushfire Recovery Program to assist environmental, community and infrastructure recovery. The programme was jointly managed by Parks Victoria and the Department of Sustainability and Environment, and recognised that a greater understanding of how fire management impacted Aboriginal cultural heritage was needed, so that heritage management practices could be improved by fire-fighting authorities. Perspectives Heritage Solutions Pty Ltd were commissioned to investigate 14 fire-affected areas in the High Country of northeast Victoria and in Gippsland (Fresløv *et al.* 2004: 1). The project brief asked the investigators to:

1. Assess and document the nature and extent of damage caused by wildfire and associated fire-suppression activities to Aboriginal cultural heritage sites across the project area, and to provide recommendations for site protection and future management.
2. Work with the Aboriginal community in documenting matters such as cultural significance and traditional interpretations of recorded sites, and in developing appropriate management options.

The resulting Post Wildfire Indigenous Heritage Survey (Fresløv 2004; Fresløv *et al.* 2004) provides a comprehensive appraisal of Aboriginal cultural heritage values in the fire-affected areas, along with an assessment of how radiant heat and fire-suppression activities impacted those sites. The assessment included fire impacts on stone artefact scatters, scarred trees and rock shelters, as well as on oral history ('intangible' or 'story place') Aboriginal heritage sites and values.

Fresløv (2004: 10) divided fire impacts into short-term and long-term impacts. Radiant heat generated by fires and fire-suppression activities were deemed short-term impacts, with post-fire erosion and ground-disturbing rehabilitation works recognised as long-term impacts (Figure 13.1). In agreement with Kelly (2003), Fresløv (2004: 16) found that low-intensity fires had minimal or no impacts on surface stone artefacts, whereas high-intensity fires caused considerable damage such as potlidding (see Chapter 7) and increased brittleness on stone artefacts. For sub-surface cultural materials, Fresløv (2004: 10) observed that sediments provided insulation from radiant heat, so that subsurface materials remained largely unaffected by the fires regardless of heat-intensity.

Figure 13.1. Hillslope near Buchan, East Gippsland, taken immediately after the 2019–2020 wildfires. Hillslopes denuded by wildfire are prone to sheet (colluvial) erosion, wind-blown erosion of sediments, and alluvial erosion of lower elevations when rivers and creeks flood. This further results in the clogging of creeks and rivers with ash and sediment for many months after the fires. Photograph by Shamis Law, courtesy of Matt Holland.

Impacts to cultural heritage values caused by fire-suppression activities and rehabilitation works, including mechanical clearing, chainsaw tree-felling, heavy-vehicle traffic, access works, track construction and road-widening, were also investigated (Fresløv 2004: 10–11). The investigation found that heavy-vehicle traffic and the construction of mineral earth containment lines damaged surface sites by fragmenting and dispersing artefacts. Of great concern, it also found that sub-surface sites unaffected by radiant heat were being damaged or destroyed by mechanical fire-suppression activities.

The major problem identified by Fresløv (2004: 25) is that Aboriginal archaeological sites are a finite resource, and that once cultural sites have been compromised or damaged by physical impacts, then their integrity can never again be restored. The essential message for the future management of GunaiKurnai sites in fire-affected areas is that each site is part of a GunaiKurnai cultural landscape that encapsulates ancestral knowledge, contemporary Aboriginal culture, and information on cultural history. From a management perspective, damage caused to a particular site is not limited to that individual locality, for all impacts to cultural sites cause irreversible, incremental damage to the broader GunaiKurnai cultural landscape and to Aboriginal culture.

In addition, GunaiKurnai sites are also irreplaceable sources of information on GunaiKurnai history—as noted above, the sites are GunaiKurnai history books written in the landscape—and that information is permanently lost when the spatial and/or stratigraphic integrity of a site is disturbed.

During the course of the Fresløv (2004: 20) Post Wildfire Indigenous Heritage Survey, 319 Aboriginal sites were identified. Most were stone artefact scatters (76%) located in high-intensity burn areas, where ground-surface visibility was correspondingly high (Freslov *et al.* 2004: 171). The construction of fire-control lines damaged 43 sites (13%), and access tracks damaged a further 57 sites (17%) (Fresløv 2004: 20). To avoid damage to GunaiKurnai cultural heritage during future fire events, Fresløv (2004: 27–31) advocated the adoption of a set of Cultural Heritage Management Principles. These Management Principles addressed three fire-event phases: 1) fire preparedness; 2) fire-suppression; and 3) fire recovery. The key aim of the Management Principles was to identify means of avoiding damage to Aboriginal cultural heritage during future fire events, including during immediate post-fire activities such as the removal of fire-damaged vegetation and infrastructure. A critical factor advocated to achieve the key management aim was the maintenance of a consultation process with Aboriginal communities during each of the three fire management phases.

The Timbarra Fire Complex wildfire (Timbarra, Sunny Point and Buchan South) started on 23 September 2017 and burnt 8963ha. The response to the fire provides an insight into how Fresløv's (2004) Management Principles were applied 13 years after they were first recommended. As part of the fire recovery strategy, the Department of Environment, Land, Water and Planning (DELWP) and Parks Victoria arranged a rapid six-day assessment by the Bushfire (Rural and Regional Affairs and Transport) team in 2017. Team deployment was initiated by the DELWP Incident Management Team on behalf of the State of Victoria. The purpose of the deployment was to prepare a report for land managers with a 'snapshot' of the priority risks posed by the fire event.

The Bushfire RRAT Report (Department of Primary Industries and Parks Victoria 2017) acknowledges Traditional Owners as one of the land management agencies, and identifies potential damage to 'Aboriginal heritage and loss of cultural sites and artefacts' as one of the top priority risks to be addressed. The report also acknowledges the GunaiKurnai Land and Waters Aboriginal Corporation (GKLaWAC) as the Registered Aboriginal Party (RAP) for the fire-affected area and alerts to reputational risks that could be caused if GKLaWAC were not provided with an opportunity to be involved in fire management. The report identifies potential risks to unknown Aboriginal sites inside fire-impacted areas, recommending:

1. That engagement of stakeholders during rehabilitation and recovery represents a valuable opportunity for land management agencies to strengthen important relationships as well as improving knowledge and understanding of heritage values, and to ensure their effective protection and preservation. Of primary importance is to maintain an open dialogue and ensure continual engagement and involvement in recovery activities.
2. That a cultural heritage evaluation of disturbed areas within the fire-affected area is undertaken with the Traditional Owners to locate, record and interpret Aboriginal sites that may have been revealed through the reduction of vegetation cover or

> exposed because of fire-suppression activities. It is important to note here that once a fire emergency is downgraded from a Level 3 incident, the *Aboriginal Heritage Act* 2006 comes into effect again; activities performed during rehabilitation and recovery phases may damage unharmed, or further damage partially harmed, sites.

The report's recommendations align well with the Management Principles advocated by Fresløv in 2004. However, it makes no reference to consultation having occurred with Traditional Owners. The report states that information regarding cultural heritage was sought from Aboriginal Affairs Victoria (then 'Aboriginal Victoria' ('AV') and now 'First Nations-State Relations'), but there is no specific record of consultation with GKLaWAC at any stage within the report.

This apparent lack of consultation is highly problematic, particularly given that the report acknowledges that there had been no comprehensive Aboriginal cultural heritage surveys of the fire area and that unknown Aboriginal sites may be present. Instead of consultation, the report relies on a desktop study citing the Aboriginal Cultural Heritage Register and Information System (ACHRIS), Aboriginal Victoria's Aboriginal sites register, which led to the conclusion that 'sites of Aboriginal heritage are known to be within the landscape but outside the fire area' (Department of Primary Industries and Parks Victoria 2017: 41). This limited reference to, and recognition of, Aboriginal heritage *outside* the fire area suggests that it is time for Fresløv's (2004) Management Principles to be re-addressed, to ensure that appropriate consultation and actions are taken to properly address the unregistered sites, which form the vast majority of Aboriginal sites across Victoria, GunaiKurnai Country included. In addition, future reporting should take due care to correctly cite agency sources (then AV, rather than AAV) and respond to appropriate legal responsibilities and obligations. The report cites legislation, regulations and codes of practice including the *Heritage Places Act* 1993 which applies to cultural heritage management practices in South Australia. Future surveys and reporting should respond to, and consider, legal responsibilities and obligations as required under the *Aboriginal Heritage Act* 2006 and the *Aboriginal Heritage Regulations* 2018.

In July and August 2019, Skelly *et al.* (2019) surveyed 64 linear kilometres of fire-affected areas on foot, and 1116km by slow-moving vehicle, following the Nunnett-Timbarra Rivers and Holey Plains wildfires that burnt 32,681ha of GunaiKurnai Country. The cultural heritage field surveys targeted areas impacted by high-intensity fires and landscapes with high cultural heritage sensitivity. Particular attention was given to areas where ground-disturbing mechanical fire-suppression and clean-up activities had been employed during and shortly after the fires, to determine how such activities may have impacted Aboriginal sites.

The surveys resulted in the identification and documentation of 99 Aboriginal sites (Holey Plains State Park area = 20 sites, Nunnett-Timbarra River area = 79 sites). Of these, 73 sites had been disturbed or destroyed. The construction of access roads and tracks was the main cause of disturbance, impacting 59 (60%) of the identified sites. Forty-eight (48%) of the sites had been disturbed by fire-suppression activities. These activities included heavy-vehicle traffic, road-widening, the construction of earthworks, and the cutting of containment lines.

Impacts to Aboriginal sites in the Nunnett-Timbarra River area were amplified by the placement of roads, tracks and containment lines on low-gradient ridgelines and spurs that are precisely the topographic features identified by GKLaWAC RAP crews as pathways followed by GunaiKurnai ancestors as they moved across the landscape. In addition, heavy-vehicle disturbance exposed artefacts formerly hidden beneath shallow layers of sediment, revealing and further heightening the presence of Aboriginal sites now in disturbed contexts. These survey findings are consistent with Frésløv (2004: 22), who had previously concluded that fire-related damage to Aboriginal sites was most likely to occur 'where control lines are placed on flat areas where flat land is at a premium on flatter ridgelines, river flats, terraces, gentle spurs above rivers'.

Legal responsibilities and obligations for the protection of Aboriginal cultural heritage are legislated under the *Aboriginal Heritage Act* 2006 and the *Aboriginal Heritage Regulations* 2018 which give effect to the Act. In the event of emergencies such as in the case of wildfires, the *Emergency Management Act* 2013 takes over. That Act defines an emergency as the 'actual or imminent occurrence of an event which in any way endangers or threatens to endanger the safety or health of any person in Victoria or which destroys or damages, or threatens to destroy or damage, any property in Victoria or endangers or threatens to endanger the environment or an element of the environment in Victoria'. The *Emergency Management Act* 2013 overrides the requirements of the *Aboriginal Heritage Act* 2006.

In relation to Aboriginal heritage management in the event of a wildfire, Aboriginal Victoria Technical Specialists and DELWP Values Officers in the Planning Section of the Incident Control Centre provide advice to the Incident Controller. This advice provides options for managing Aboriginal cultural heritage. It is important to note for such emergency planning, that when ACHRIS managed by Aboriginal Victoria shows a landscape without any registered Aboriginal Places, that absence of registered Aboriginal Places does not necessarily mean their absence in that landscape. Rather, it usually simply means that none are known and registered because appropriate cultural heritage surveys have not yet been undertaken to generate the necessary data. Skelly *et al.* (2019) emphasised that, consistent with international standards and legislation for the protection of Indigenous cultural heritage, the 'any property' that the *Emergency Management Act* 2013 specifically aims to protect should include cultural heritage (Aboriginal) sites such as those identified in this volume.

Frésløv's (2004: 28–30) previously developed set of management principles based on cultural heritage surveys following wildfires impacting large parts of East Gippsland in 2003 (see above for details) are as relevant today as when first formulated:

1. *Fire prevention planning and preparedness.* Planning should take place in collaboration with Aboriginal Traditional Owners to facilitate the protection of Aboriginal cultural heritage sites. This approach should re-affirm the principles of the state-wide Code of Practice for fire management on public land, improve consultation and communication with Aboriginal communities and increase awareness of Aboriginal cultural heritage values among fire managers and fire-suppression personnel, thus providing the highest possible protection for Aboriginal cultural heritage sites during fire-suppression activities.
2. *Fire-suppression.* During any fire event, Aboriginal cultural heritage sites should be protected and managed in a cooperative, strategic and sensitive way in consultation with affected Aboriginal communities.

3. *Fire rehabilitation and recovery*. Following fire events, impacts to Aboriginal cultural heritage sites should be identified through community consultation and cultural heritage surveys in fire-affected areas. Fire rehabilitation and recovery plans should be developed in collaboration with Aboriginal communities to minimise impacts to Aboriginal heritage sites. Landscape rehabilitation following a fire event can itself damage or destroy cultural heritage sites. Therefore, prior to landscape rehabilitation, the fire rehabilitation and recovery plan needs to incorporate an Aboriginal site management plan.

Key to the success of proposed activities and actions is ongoing consultation with the local Aboriginal communities during fire prevention, fire-suppression and fire recovery.

Chapter 14

Understanding the Distribution and Impacts of Wildfires in GunaiKurnai Country through Subregions

Jessie Buettel, Stefania Ondei, Bruno David, Joanna Fresløv and Russell Mullett

GunaiKurnai Country contains a wide range of environmental zones. Differentiating the study region by environmental zone or 'bioregion' allows for a more detailed understanding of the spread of wildfires and the distribution of cultural sites relative to vegetation, landforms, infrastructure and centres of population. Such an approach allows us to make better predictions on the actual and likely impacts of fires on cultural sites across the landscape. While a fire may begin in one bioregion, it may (or may not) spread to another: understanding the spread of fires within and across bioregions, each with its own fuel loads and distribution of cultural sites, may shed useful light on management needs or options at a landscape scale.

Here we have used the national *Interim Biogeographic Regionalisation of Australia* (IBRA) bioregion framework to sub-divide the GunaiKurnai Land and Waters Aboriginal Corporation Registered Aboriginal Party (GKLaWAC RAP) and nearby areas into analytical landscape units. There are 89 major bioregions across Australia. These can be sub-divided into 419 smaller subregions

Table 14.1. The IBRA subregions (Australia's bioregions (IBRA)—The National Reserve System (NRS) | Department of Agriculture, Water and the Environment) that intersect and sometimes border the GKLaWAC RAP area. Shown is the broader bioregion each subregion is within, the total area of each subregion (in km^2) and the percentage of the subregion area that is within the GKLaWAC RAP area.

Name	Code	Bioregion	Area (km^2)	% in GKLaWAC Country
Snowy Mountains	AUA01	Australian Alps	7131.1	2.20%
Victorian Alps	AUA02	Australian Alps	5198.7	42.24%
Wilsons Promontory	FUR01	Furneaux	411.4	0.04%
Gippsland Plain	SCP01	South East Coastal Plain	12470.0	58.87%
East Gippsland Lowlands	SEC01	South East Corner	6235.2	30.38%
South East Coastal Ranges	SEC02	South East Corner	17345.2	17.61%
Highlands—Southern Fall	SEH01	South Eastern Highlands	11963.3	63.45%
Highlands—Northern Fall	SEH02	South Eastern Highlands	14158.1	0.74%
Strzelecki Ranges	SEH04	South Eastern Highlands	3420.5	57.13%
Kybeyan Gourock	SEH15	South Eastern Highlands	4792.2	0.00%
Monaro	SEH16	South Eastern Highlands	12675.4	0.05%

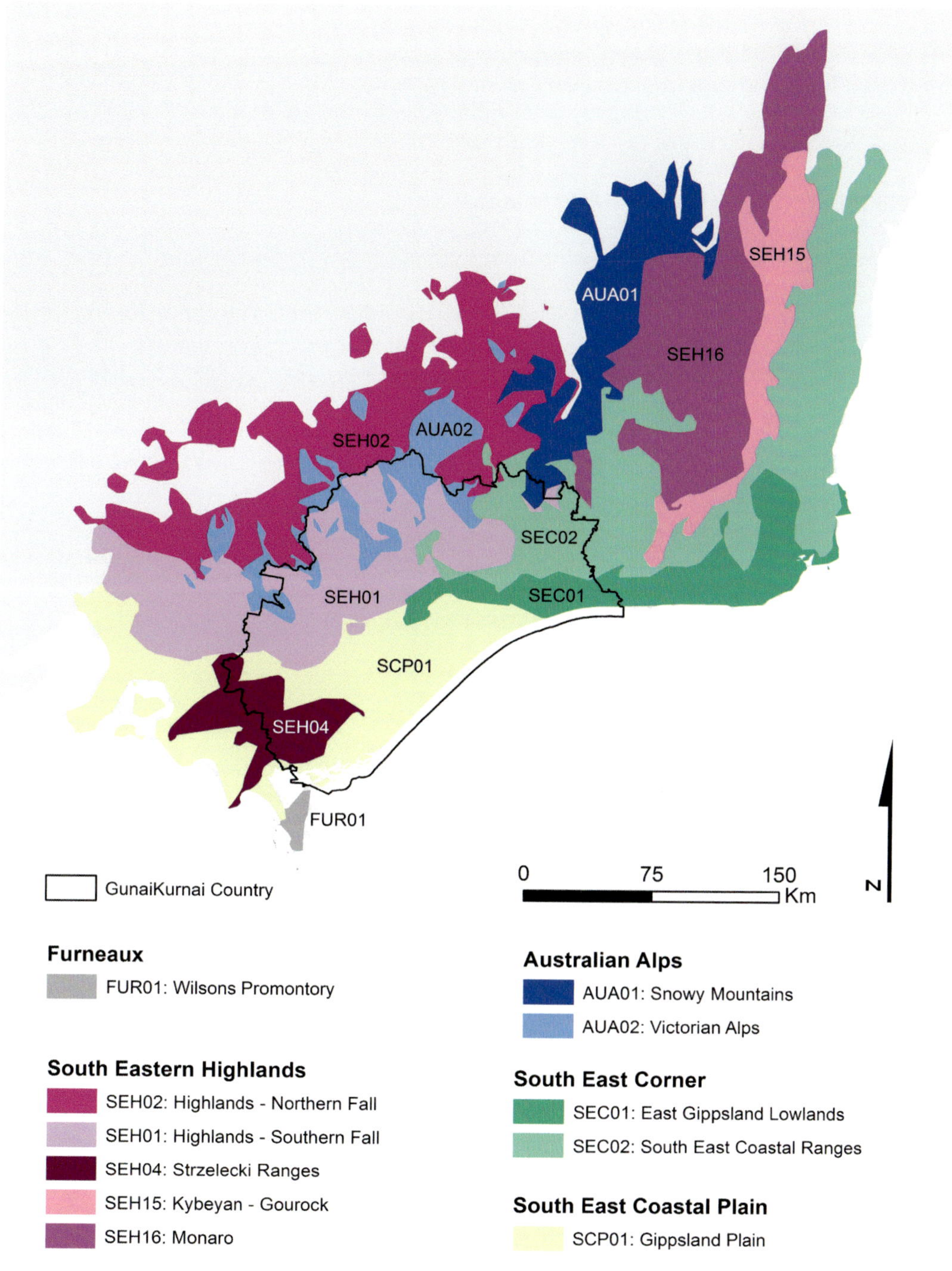

Figure 14.1. Location and extent of each IBRA subregion (borders are shown simplified for clarity) in and surrounding the GKLaWAC RAP area (black border line). Each subregion is grouped and colour-coded by the broader IBRA bioregion that it is part of; these bioregions are listed in bold in the legend.

that consist of more localised land systems within the bioregions. Each of these bioregions and subregions is environmentally and climatically distinct, and grouped according to common climate, geology, landform, native vegetation, and species information. For this chapter, we are interested in the IBRA subregions that intersect with or are in close proximity to the GKLaWAC RAP area. There are 11 subregions that fit this criterion (Figure 14.1, Table 14.1).

THE SUBREGIONS OF THE GKLAWAC RAP AREA

Each subregion has a distinct ecology, biodiversity, and climate (Figure 14.1, Tables 14.2–14.4).

AUA01 Snowy Mountains. Located towards the northeast and intersecting a small area of the GKLaWAC RAP area, this is a mountainous environment that reaches an average elevation of 1315m above sea level (a.s.l.), with Eucalypt Forests and Eucalypt Woodland dominating the vegetation. Much (71%) of this subregion is under conservation management (see Chapter 6 for 'conservation' management as a land use type), and 27% is under minimal human development. This is also the only subregion in the GKLaWAC RAP area to have extensive areas of Grassland (12%).

Table 14.2. Percentage of area each vegetation type contributes to extent of each subregion. All 11 subregions that sit within and surrounding the GKLaWAC RAP area are shown. All proportions greater than 30% have been shaded to highlight the vegetation type and its larger contribution to the area within that subregion.

Subregion	Cleared	Other	Grassland	Shrubland	Eucalypt Woodland	Eucalypt Forest		Rainforest
						Dry Sclerophyll	Wet Sclerophyll	
Snowy Mountains	1	0	12	1	34	29	21	0
Victorian Alps	0	2	1	3	54	1	38	1
Wilsons Promontory	0	2	1	37	18	28	10	3
Gippsland Plain	71	1	1	6	9	4	3	0
East Gippsland Lowlands	11	7	0	3	10	17	50	0
South East Coastal Ranges	14	3	0	1	22	30	28	1
Highlands—Southern Fall	13	0	0	2	4	20	61	1
Highlands—Northern Fall	14	0	0	0	5	27	52	0
Strzelecki Ranges	69	0	0	0	1	1	28	1
Kybeyan Gourock	27	0	0	1	14	11	45	1
Monaro	64	0	1	0	9	22	3	0

Table 14.3. Percentage of area each land-use type contributes to each subregion. All 11 subregions that sit within and surrounding the GKLaWAC RAP area are shown. Land-use type is in order of increasing intensity of human-use, from most intensive uses on the left to least intensive on the right. All proportions greater than 30% have been shaded to highlight that vegetation type and its larger contribution to the area within the subregion.

Subregion	Intensive uses	Production from agriculture and plantations	Production from relatively natural environments	Nature conservation	Water
Snowy Mountains	0	1	27	71	1
Victorian Alps	2	1	48	49	0
Wilsons Promontory	0	0	0	100	0
Gippsland Plain	22	61	3	13	1
East Gippsland Lowlands	3	15	56	26	0
South East Coastal Ranges	1	14	44	41	1
Highlands—Southern Fall	10	14	50	25	0
Highlands—Northern Fall	3	22	54	21	0
Strzelecki Ranges	11	84	1	4	0
Kybeyan Gourock	0	12	57	30	0
Monaro	4	32	52	9	4

Table 14.4. Average elevation, mean annual temperature (MAT) and mean annual precipitation (MAP) for each subregion within and surrounding the GKLaWAC RAP area. The means (average) and minimum and maximum values are shown for both MAT and MAP, to show variability.

Subregion	Mean annual temperature (°C)			Elevation (m a.s.l.)	Mean annual precipitation (mm)		
	MIN.	MAX.	MEAN	MEAN	MIN.	MAX.	MEAN
Snowy Mountains	3.30	13.10	8.42	1315.775	570	2377	1254
Victorian Alps	5.30	12.40	8.89	1167.457	689	2308	1382
Wilsons Promontory	10.90	14.60	13.73	151.778	811	1438	1020
Gippsland Plain	12.90	14.90	14.32	47.881	561	1094	746
East Gippsland Lowlands	10.60	15.50	14.35	138.992	623	1168	910
South East Coastal Ranges	8.70	15.90	13.35	400.052	512	1300	845
Highlands—Southern Fall	7.20	14.70	12.39	449.673	602	1692	980
Highlands—Northern Fall	8.00	14.50	11.90	670.235	618	1673	1117
Strzelecki Ranges	10.70	14.40	13.39	209.872	717	1428	1037
Kybeyan Gourock	8.00	14.10	10.80	931.133	583	1428	840
Monaro	6.40	13.40	11.00	889.619	488	1640	653

AUA02 Victorian Alps. Also often referred to as the 'High Country', this subregion is located in the north of GunaiKurnai Country and is widely recognised for its extensive mountain system that in the GKLaWAC RAP area reaches a maximum peak of 1986m a.s.l. (Mount Bogong), the highest peak in Victoria. Like the Snowy Mountains, this subregion is in the Australian Alps bioregion and therefore also has a high average elevation (1167m a.s.l.). The majority of this subregion is covered by Eucalypt Woodland vegetation (54%) with intersecting Eucalypt Forest (39%) in the gullies. In the higher areas of the subregion, Shrubland and Grassland dominate, and towards its southern end there are some pockets of Rainforest. In the Victorian Alps, there is much exposed bedrock across the entire range, and almost all of the land area (97%) is under conservation management (49%) and minimal human development (48%).

FUR01 Wilsons Promontory. This is a small subregion (411km^2), consisting of a small peninsula off the southern end of the Australian Coastline, and part of the larger 'Furneaux' bioregion. Only a small area (0.04%) of this subregion intersects with the southern end of the GKLaWAC RAP area. This subregion is one of the 11 within this broad region that has a high proportion of Shrubland relative to its total area. It contains a mix of Eucalypt Forests (28% Dry Sclerophyll and 10% Wet Sclerophyll), Shrubland (37%) and Eucalypt Woodland (18%). Of note, a pocket of Eucalypt Forest—Wet Sclerophyll is found in the centre of this subregion, with small patches of Rainforest also surrounding and intersecting it. All of this subregion is under conservation management.

SCP01 Gippsland Plain. The Gippsland Plain ranks second among these 11 subregions (behind the Strzelecki Ranges) as having the most extensively modified subregion—83% of the area is under the two highest categories of human use (intensive uses 22% and production from agriculture and plantations 61%). It also has a high overlap with the GKLaWAC RAP area. This subregion is very flat, with an average elevation of 48m a.s.l., making it well suited for recent urban and rural developments (agriculture and farming).

SEC01 East Gippsland Lowlands. This subregion is located along the southeast coastline of GunaiKurnai Country. It consists of large tracts of Eucalypt Forest—Wet Sclerophyll intersected with pockets of Eucalypt Forest—Dry Sclerophyll throughout. There are also large areas of Cleared land, but these are largely settlements and urban infrastructure rather than farming. Much of the land-use (56%) is in fact production in areas that have seen little infrastructure development and that retain uncleared or sparsely cleared land.

SEC02 South East Coastal Ranges. This subregion has among the highest proportion of Eucalypt Forest—Dry Sclerophyll (30%), but is actually also very mixed with an equally high percentage of both Eucalypt Woodland and Eucalypt Forest—Wet Sclerophyll. There is also a moderately high percentage of Cleared land (12%), which is mostly found in the northeastern part of the subregion.

SEH01 Highlands—Southern Fall. This subregion has the highest amount of overlap with the GKLaWAC RAP area (63%) of all the 11 sub-regions included in this chapter. It also has some of the highest coverage of Eucalypt Forest—Wet Sclerophyll (61%). Much of the Cleared area is towards the western half of the subregion, with mostly Eucalypt Forest—Dry Sclerophyll intersecting with these areas.

SEH02 Highlands—Northern Fall. Surrounds and hugs the northern part of the GKLaWAC RAP area, with only a small area intersecting it (0.7%). It consists of a mixture of Dry Sclerophyll and Wet Sclerophyll Eucalypt Forests, with the wetter areas extending towards the southwestern end of the subregion, where there are also tiny pockets of Rainforest (which is found nowhere else within the subregion).

SEH04 Strzelecki Ranges. Has the highest percentage (95%) of area used for human production—agriculture and plantations (84%) and intensive uses (11%). Although much of this subregion is heavily human-impacted, there remains a moderate amount of Eucalypt Forest—Wet Sclerophyll located primarily on the eastern gullies and ridges, where there is higher rainfall and more complex topography.

SEH15 Kybeyan Gourock. Has a high degree of low-level human use (production from relatively natural environments 57%) and a high percentage of nature conservation areas (30%). Of the vegetation types, Eucalypt Forest—Wet Sclerophyll are the most dominant (45%), followed by a relatively high amount of Cleared land (27%).

SEH16 Monaro. Much of the interior of this subregion is Cleared, with 32% used for production from agriculture and plantations, and 52% for production from relatively natural environments. There is a small section of Eucalypt Woodland in the southern end of the subregion (which is the end closest to, and bordering on, the GKLaWAC RAP area boundary), and a small amount of Grassland towards its northern tip. Besides the Cleared land, the dominant vegetation type is Eucalypt Forest—Dry Sclerophyll (22%), located mostly around the edges of the subregion.

IN WHICH SUBREGIONS HAVE SURVEYS FOR CULTURAL SITES BEEN UNDERTAKEN IN THE GKLAWAC RAP AREA?

The vast majority of registered cultural sites have been identified and reported through systematic archaeological surveys undertaken for cultural heritage management assessments, although occasionally sites were individually registered by interested parties. These surveys have targeted delimited areas planned for development, such as road construction corridors. In most cases, the individual cultural heritage management field surveys have been small in area, sometimes only measuring hundreds of square metres. Cultural site field surveys have also been undertaken for a range of monitoring programs, such as to assess the impacts of visitors on sites in the Gippsland Lakes region (e.g., Fresløv 2022), or the distribution of sites in delimited areas following wildfires (e.g., David *et al.* 1998; Fresløv 2004; Skelly *et al.* 2019). Given the very selective, targeted survey locations, any extrapolation of known (registered) site distributions to the broader landscape needs to take the location and extent of the surveyed areas into account. However, such data of where the field surveys have been done across the GKLaWAC RAP area have never been compiled. Those data about individual field surveys (where reported, which is not the case for every field survey undertaken) lie in hundreds of unpublished reports. These field surveys varied enormously in quality. They sometimes consist of carefully undertaken, systematic field walking involving closely-spaced (2–5m apart) lines of field walkers; sometimes broadly-spaced field walkers spaced 10m apart or more; sometimes vehicle surveys where the surveyor(s) slowly drove along delimited transect areas or vehicle tracks without leaving their cars; and all these options were undertaken across terrains that could range from excellent (100%) to very poor (down to 0%) conditions of ground visibility. Assessing the geographical distribution, density and quality of the field surveys that have been undertaken since the 1970s, and which have resulted in the registration of the vast majority of cultural sites currently in the VAHR, is beyond this chapter, and is recommended for future research.

THE DISTRIBUTION OF CULTURAL SITES IN THE SUBREGIONS OF THE GKLAWAC RAP AREA

The subregions that take up the largest parts of the GKLaWAC RAP area are the Gippsland Plain (7343km^2) and Highlands—Southern Fall (Figure 14.2). Around 72% (n = 1931) of all registered cultural sites are found within the Gippsland Plain subregion, compared to 8% (n = 213) in Highlands—Southern Fall, despite both covering approximately the same area of the GKLaWAC RAP area (Figure 14.2). This may indicate that the Gippsland Plain subregion within the GKLaWAC RAP area has been more intensely surveyed than the other subregions, although the data on exactly where all the field surveys have been undertaken in GunaiKurnai Country are not currently available. Conversely, the Highlands—Southern Fall has few registered cultural sites for the larger part of the GKLaWAC RAP area that it occurs in.

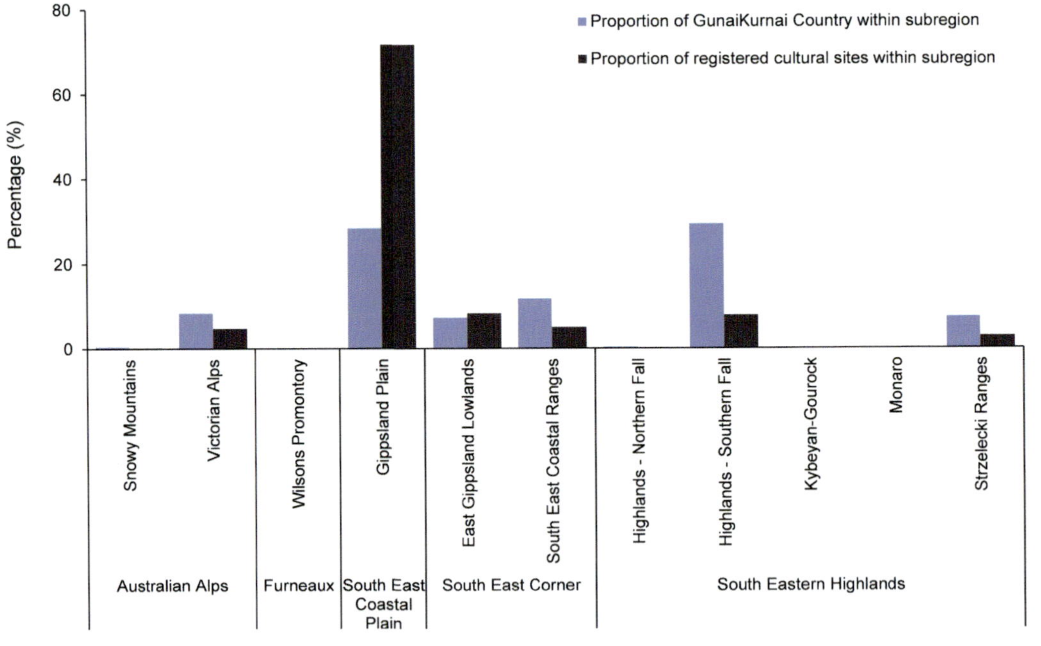

Figure 14.2. Percentage of total area of the GKLaWAC RAP area located within each subregion (purple bars) and proportion of total number of registered cultural sites (n = 2698) within each subregion (dark grey bars), as defined by the boundaries of the GKLaWAC RAP area. See Figure 14.1 for a graphic representation of the subregions and how much area each covers within the GKLaWAC RAP area.

FIRE HISTORY AND SUBREGIONS IN SE AUSTRALIA

Here we present, for each individual subregion:

1. Area burnt by wildfires and prescribed burning in the 70 years between 1930 and 1999, and in the 21 years between 2000 and 2020.
2. Area burnt by wildfires and prescribed burning by decade.
3. A closer look at the extent of area burnt by wildfires and prescribed burning between 2000 and 2020 (the past 21 years).
4. The extent of each vegetation type present in each subregion, and area burnt by vegetation type since 2000.

For each plot of area burnt through time, we present both the non-cumulative total area and the cumulative total area burnt. In other words, we present:

1. The total (non-cumulative) area burnt in each century or decade, with each burnt area only counted once; and
2. The total area burnt per year, summed for each time category (decade or century). If an area was burnt multiple times in different decades or centuries, then it was counted those multiple times.

Unless otherwise stated, all areas and percentages given in-text to describe the patterns in fire history and extent pertain to non-cumulative areas, representing each site only once.

Snowy Mountains. The Snowy Mountains subregion has witnessed a greater area of burning through wildfires in the past 21 years (2000–2020) than in the previous 70 years (1930–1999) (Figure 14.3). Since 2000, almost the entire extent of the Snowy Mountains has been burnt by wildfires (Figure 14.5), with many areas having been burnt more than once over this time (Figure 14.3). In contrast, a greater area underwent prescribed burning in the 20th century (1567km^2) than in the 2000s (322km^2; Figure 14.3).

The decade that saw the largest area burnt by wildfires was the 2000s (4918km^2), followed by the 2010s (2856km^2) and the 1930s (2893km^2). The 2000s also saw the largest cumulative area burnt by wildfires (the same locations burnt more than once per year). There were only relatively small areas burnt by wildfires in all other decades (Figure 14.4). The area burnt by prescribed burning events was consistent across all decades (average area = 273km^2, non-cumulative), with very few areas burnt more than once by prescribed burns (Figure 14.4).

Much of the Snowy Mountains subregion (87%) has been burnt by wildfires since 2000. Much of the area burnt by wildfire is proportional to the total area of each vegetation type within the subregion (Figure 14.5). Prescribed burning has only burnt across 5% of the total extent of the subregion, and much of this occurred in the Eucalypt Woodland (2.3%, 150km^2) and Eucalypt Forest—Dry Sclerophyll (1.8%, 117km^2).

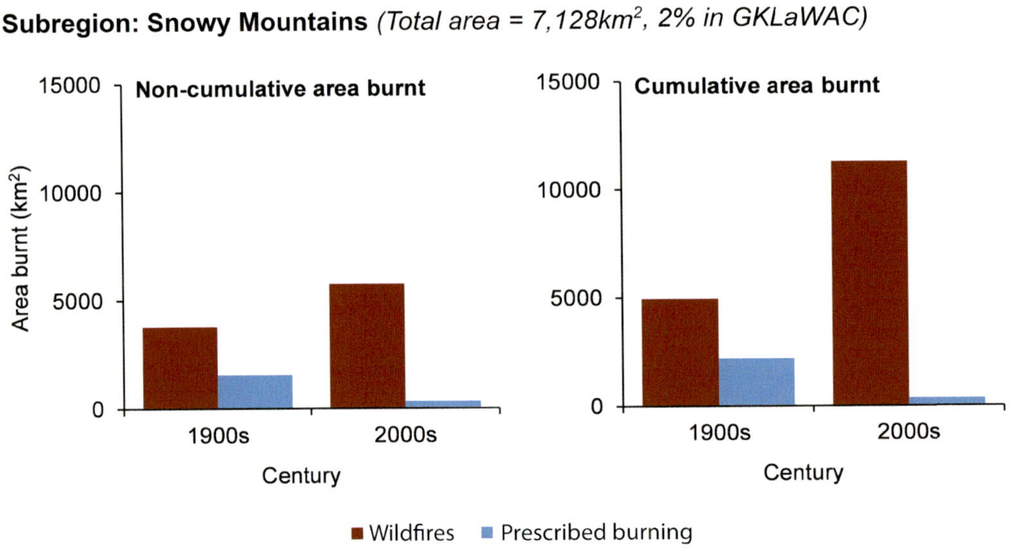

Figure 14.3. Area of the Snowy Mountains subregion burnt (km²) in the 1900s (1930–1999) and 2000s (2000–2020). Brown bars represent fires caused by wildfire events, blue bars by prescribed burns. The plot on the left shows the total non-cumulative area burnt in each century (areas might have been burnt more than once), the plot on the right the total area burnt per year, summed for each century (cumulative) and thereby accounting for areas burnt more than once over that time.

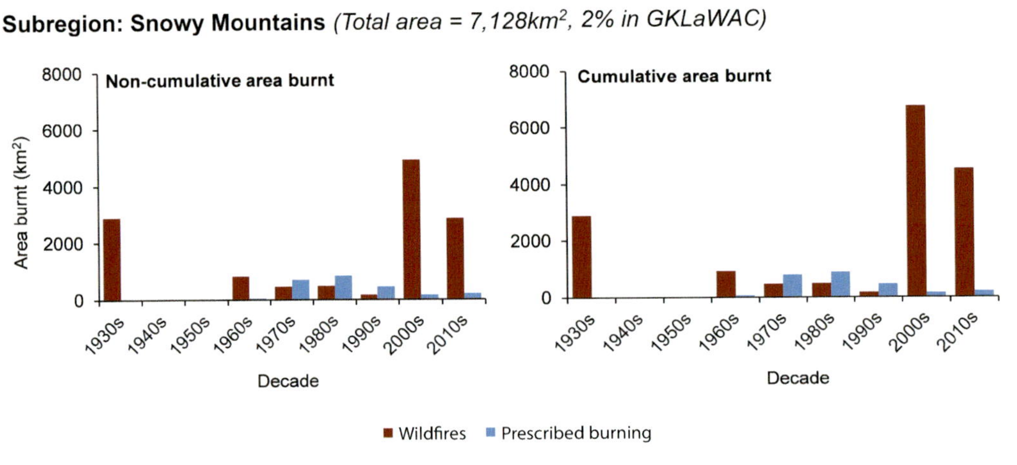

Figure 14.4. Extent of the Snowy Mountains subregion burnt (km²) in each decade from the 1930s to 2010s. Brown bars represent areas burnt by wildfires, blue bars by prescribed burns. The plot on the left shows the total non-cumulative area burnt in each decade (areas might have been burnt more than once), the plot on the right the total area burnt per year, summed for each decade (cumulative) and thereby accounting for areas burnt more than once over that time.

Australian Alps

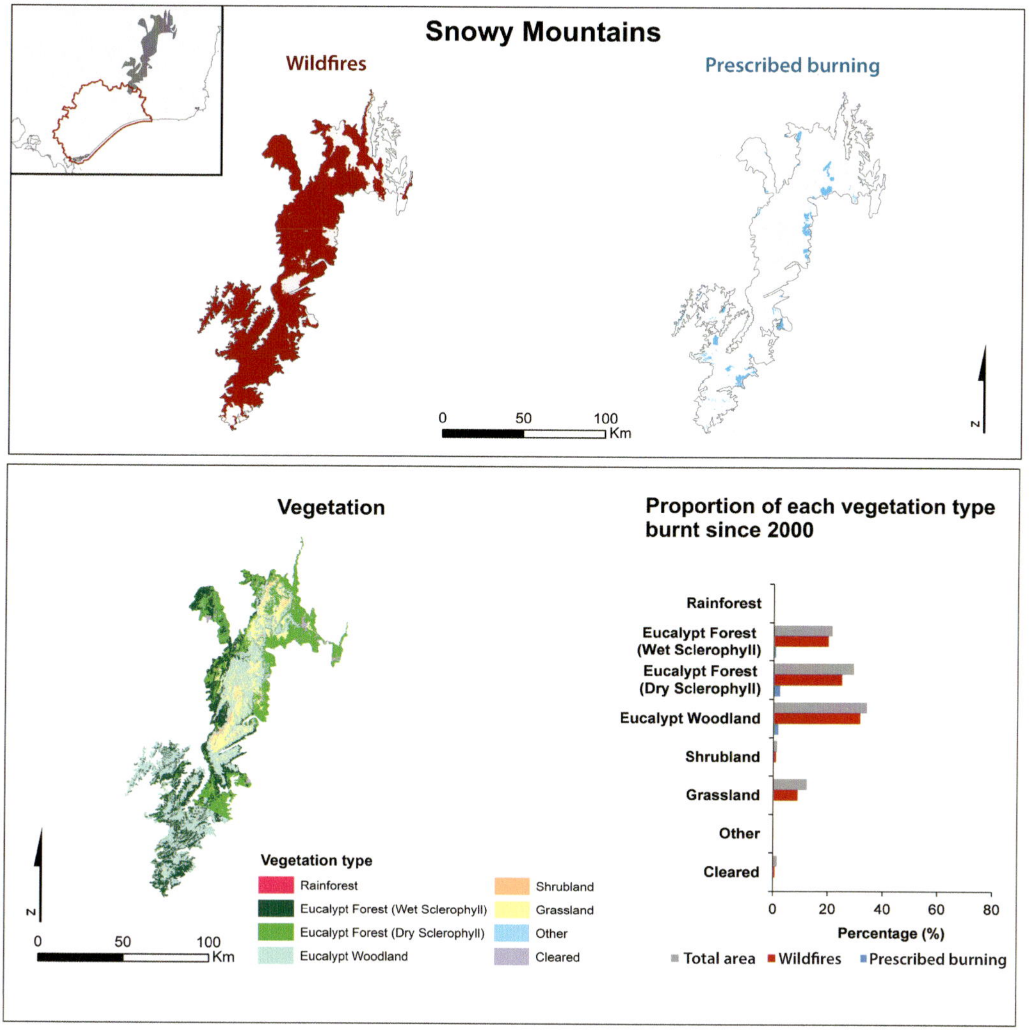

Figure 14.5. Top panel: Extent of wildfires (left, in red) and prescribed burns (right, blue) in the Snowy Mountains subregion since 2000. Bottom panel: Left: Extent of each of the different vegetation types present within the subregion. Right: Area burnt since 2000, by vegetation type.

Victorian Alps. The Victorian Alps subregion has had more non-cumulative area burnt by wildfires between 1930–1999 than between 2000–2020 (Figure 14.6). However, this pattern flips when areas burnt more than once are taken into account (cumulative area burnt): a much larger area burned in the 21 years of the 21st century than in the 70 years of the 20th (Figures 14.3, 14.6). In contrast, the area burnt by prescribed burns was higher in the 20th century (4924km^2) than in the 2000s (4434km^2; Figure 14.6).

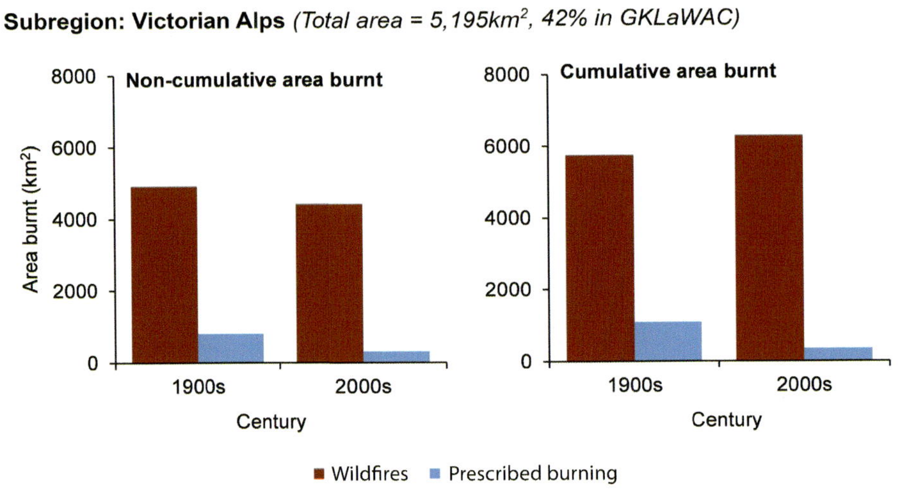

Figure 14.6. Extent of the Victorian Alps subregion burnt (km²) in the 1900s (1930–1999) and 2000s (2000–2020). Brown bars represent fires caused by wildfire events, blue bars by prescribed burns. The left plot shows the total non-cumulative area burnt in the 1900s vs. 2000s (areas might have been burnt more than once). The right plot shows the total area burnt per year, summed for each century (cumulative); this accounts for areas that might have been burnt more than once during that time.

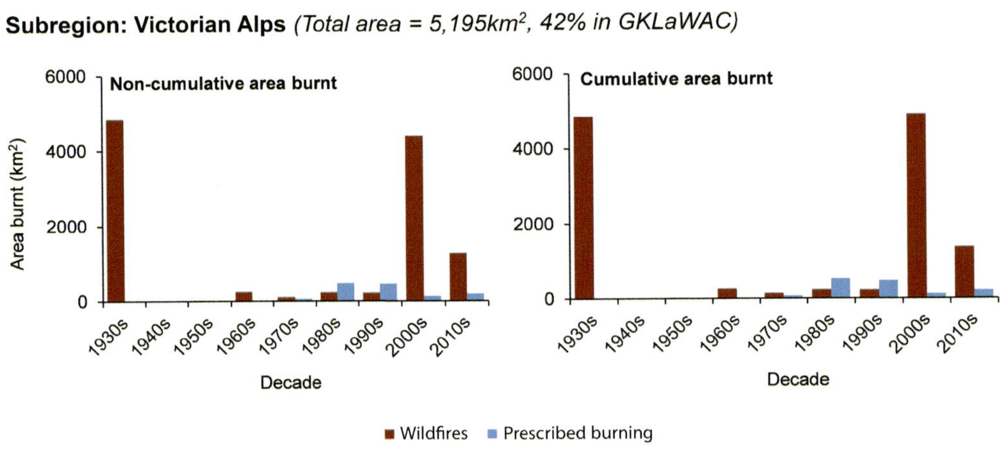

Figure 14.7. Extent of the Victorian Alps subregion burnt (km²) in each decade from the 1930s to 2010s. Brown bars represent areas burnt by wildfires, blue bars by prescribed burns. The left plot shows the total area burnt in each decade (non-cumulative: areas might have burnt more than once), and the right plot shows the total cumulative area burnt per year, summed for each decade; this accounts for areas that may have burnt more than once over that time.

The decade that saw the largest area burnt by wildfires was the 1930s (4846km²), followed by the 2000s (4399km²). There was very little difference between the non-cumulative and cumulative areas burnt, meaning that few locations burnt multiple times in each decade (Figure 14.7). Relatively few, small areas burnt by wildfires in all other decades, especially from the 1940s to 1990s (Figure 14.7). On the other hand, the area burnt by prescribed burning was double that of wildfires for the 1980s (prescribed burns: 495km^2, wildfires: 239km^2) and 1990s (prescribed burns: 467km^2, wildfires: 231km^2), with very few areas burnt by prescribed burns in the other decades (Figure 14.7).

Much of the Victorian Alps subregion (85%) has been burnt by wildfires since 2000. Eucalypt Woodland covers 54% of the total area of this subregion, and therefore has the highest percentage of area burnt by wildfires (51%). Indeed, almost all (94%) of the total area covered by Eucalypt Woodland has been burnt by wildfires, whereas this is so of only 74% of the total area covered by Eucalypt Forest—Wet Sclerophyll (Figure 14.8). Eucalypt Forest—Dry

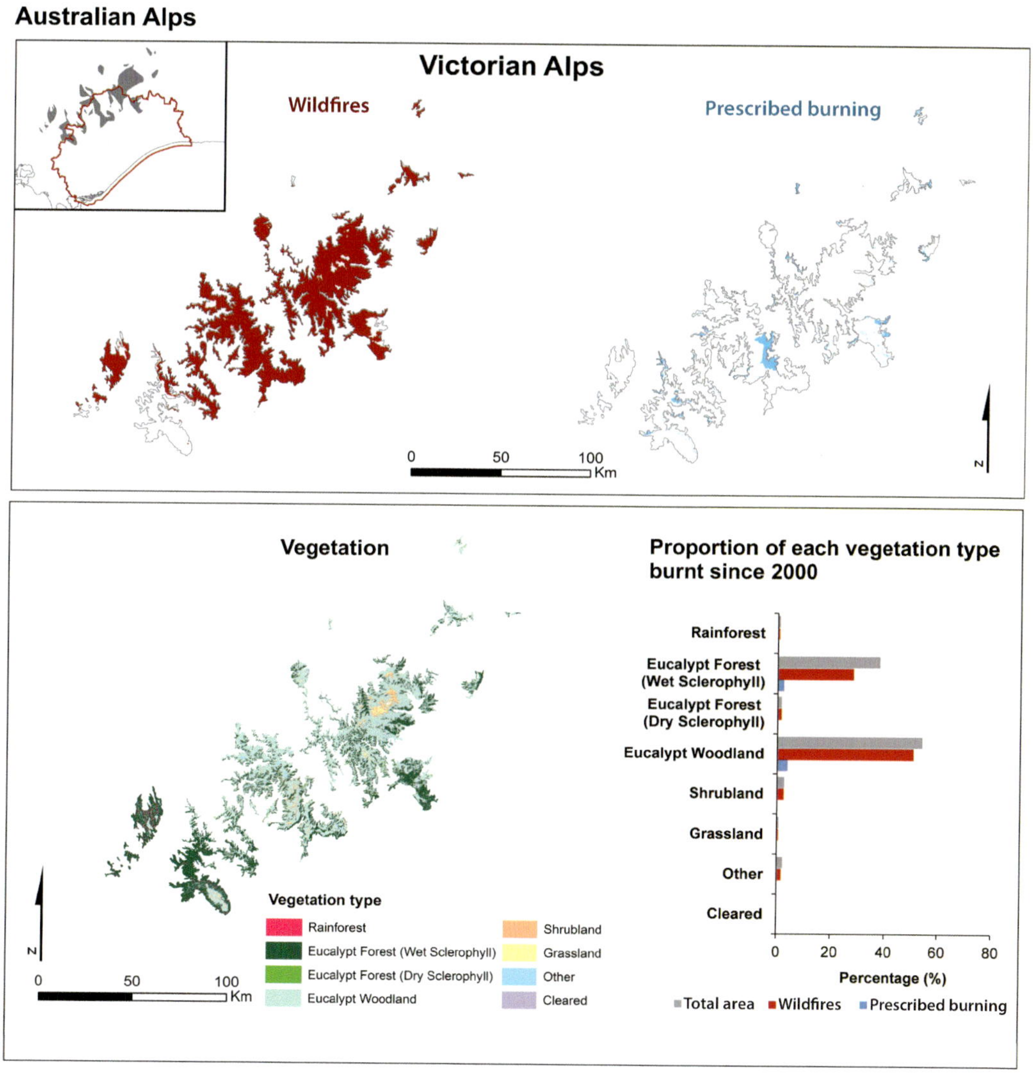

Figure 14.8. Top panel: Extent of wildfires (left, in red) and prescribed burns (right, blue) in the Victorian Alps subregion since 2000. Bottom panel: Left: Extent of each vegetation type in the subregion. Right: Graphic representation of area burnt since 2000, by vegetation type.

Sclerophyll only makes up a small area within this subregion (total = 74km^2), but of this small area, all has been burnt by wildfire since 2000 (Figure 14.8). Prescribed burning has burnt only 6.7% of the total area of the subregion, and much of this occurred in Eucalypt Woodland (3.8%, 202km^2) and Eucalypt Forest—Wet Sclerophyll (2.2%, 118km^2). A small pocket of the southwestern end of the subregion has not been burnt by wildfire, and only slightly by prescribed burning (Figure 14.8).

Wilsons Promontory. The Wilsons Promontory subregion has seen more non-cumulative area burnt by wildfires between 1930–1999 than between 2000–2020, and this pattern holds true for cumulative areas burnt also (area burnt multiple times). There was a slight increase in the amount of area burnt more than once in the 21st century compared to in the 20th century. While the total area burnt by prescribed burns was small in both centuries, it was slightly higher in the 20th century than in the 21st (Figure 14.9).

The decade that saw the largest area burnt by wildfires was the 2000s (299km^2, 73% of the total subregion area), followed by the 1950s (255km^2, 62% of the total subregion area). There was very little difference between the non-cumulative and cumulative areas burnt. That is, this subregion witnessed few locations that burnt multiple times in each decade (Figure 14.10). There were no reported wildfires for the 1930s, 1940s or 1980s, and a very small area burnt by wildfires in the 1990s and 2010s (Figure 14.10). The area burnt by prescribed fires was small across all decades, perhaps owing to the subregion's small total area (411km^2).

The Wilsons Promontory subregion covers a relatively small area, with most vegetation types covering less than 100km^2 in area. Of this small area, 74% (299km^2) has been burnt by wildfires, and 2.5% by prescribed burning (Figure 14.11). Shrubland makes up the majority of

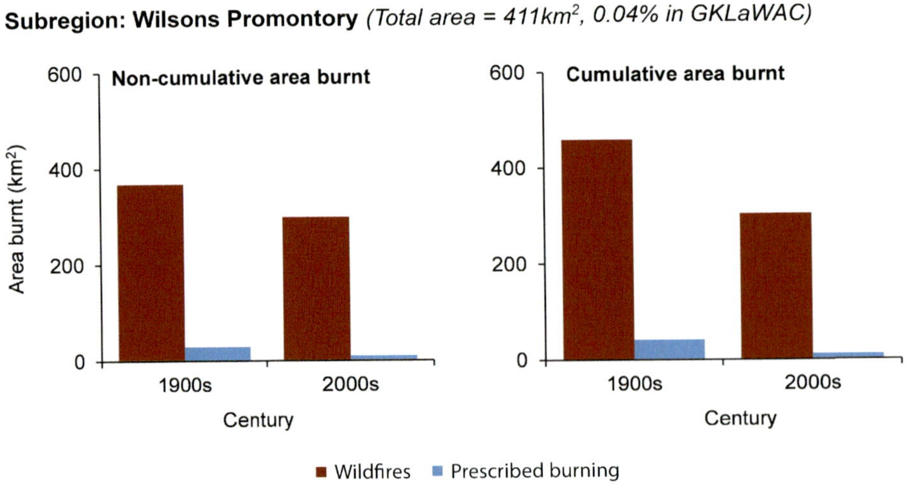

Figure 14.9. Extent of the Wilsons Promontory subregion burnt (km^2) in the 1900s (1930–1999) and 2000s (2000–2020). Brown bars represent areas burnt by wildfires, blue bars by prescribed burns. The left plot shows the total non-cumulative areas burnt in each century (areas might have been burnt more than once), the right plot total areas burnt per year, summed for each century (cumulative); this accounts for areas burnt more than once during those times.

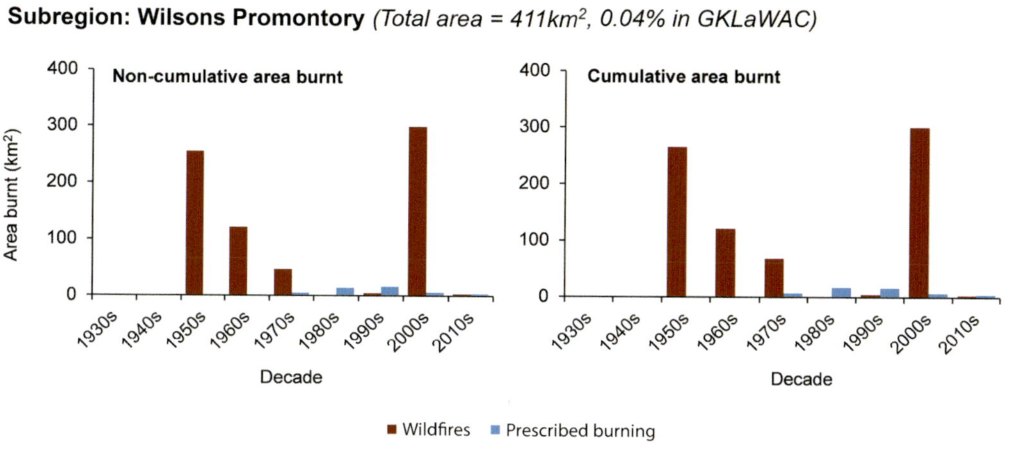

Figure 14.10. Extent of the Wilsons Promontory subregion burnt (km²) in each decade from the 1930s to 2010s. Brown bars represent wildfires, blue bars prescribed burns. The left plot shows the total area burnt in each decade (non-cumulative: areas may have burnt more than once), the right plot total cumulative areas burnt per year, summed for each decade; this accounts for areas burnt more than once over those periods.

the subregion, and thus has the highest percentage area burnt across all vegetation types. The pocket of Eucalypt Forest—Wet Sclerophyll with intersecting Rainforest towards the centre of the subregion remained largely unburnt since 2000 (Figure 14.11).

Gippsland Plain. Even though more non-cumulative area was burnt by wildfires between 1930–1999 than in the past 21 years (2000–2020), a relatively small proportion of the total area has burnt during each phase (7.5% and 4.5%, of the total area, respectively). While the total area burnt by prescribed burns was also low in both periods, it was slightly higher between 2000–2020 than in the 1900s (Figure 14.12).

Across each decade, there was very little area burnt by wildfires compared to the total area of the subregion (12,463km²). The decade that saw the largest area burnt by wildfires was the 1930s with 322km², representing 2.6% of the total area of the subregion (Figure 14.13). In other words, wildfires are less frequent in this subregion, also highlighted by the small proportion of area burnt by each vegetation type (Figure 14.14). The vegetation types that had the largest areas burnt by wildfires included Eucalypt Woodland with 162km² (1.3% of total subregion area burnt by wildfires) and Cleared vegetation with 209km² (1.7% of the total subregion area burnt by wildfires; Figure 14.14). Although the area burnt by prescribed burns surpassed the area burnt by wildfires in the 1980s and 1990s, the total area remained consistently low across all decades, with a maximum of 1.7% (reached in the 2010s) of the total non-cumulative area burnt by prescribed burns across all decades.

East Gippsland Lowlands. In both centuries (1930–1999, 2000–2020), much of the total area of the subregion has been burnt by wildfires at least once (1990s: 76% of total subregion area; 2000s: 69%), with a slightly larger area burnt in the 1900s (4712km², Figure 14.15). However, a much greater area of the East Gippsland Lowlands has burnt multiple times, particularly in the 1900s (non-cumulative area burnt: 4712km²; cumulative area burnt: 7249km²). The same

Furneaux

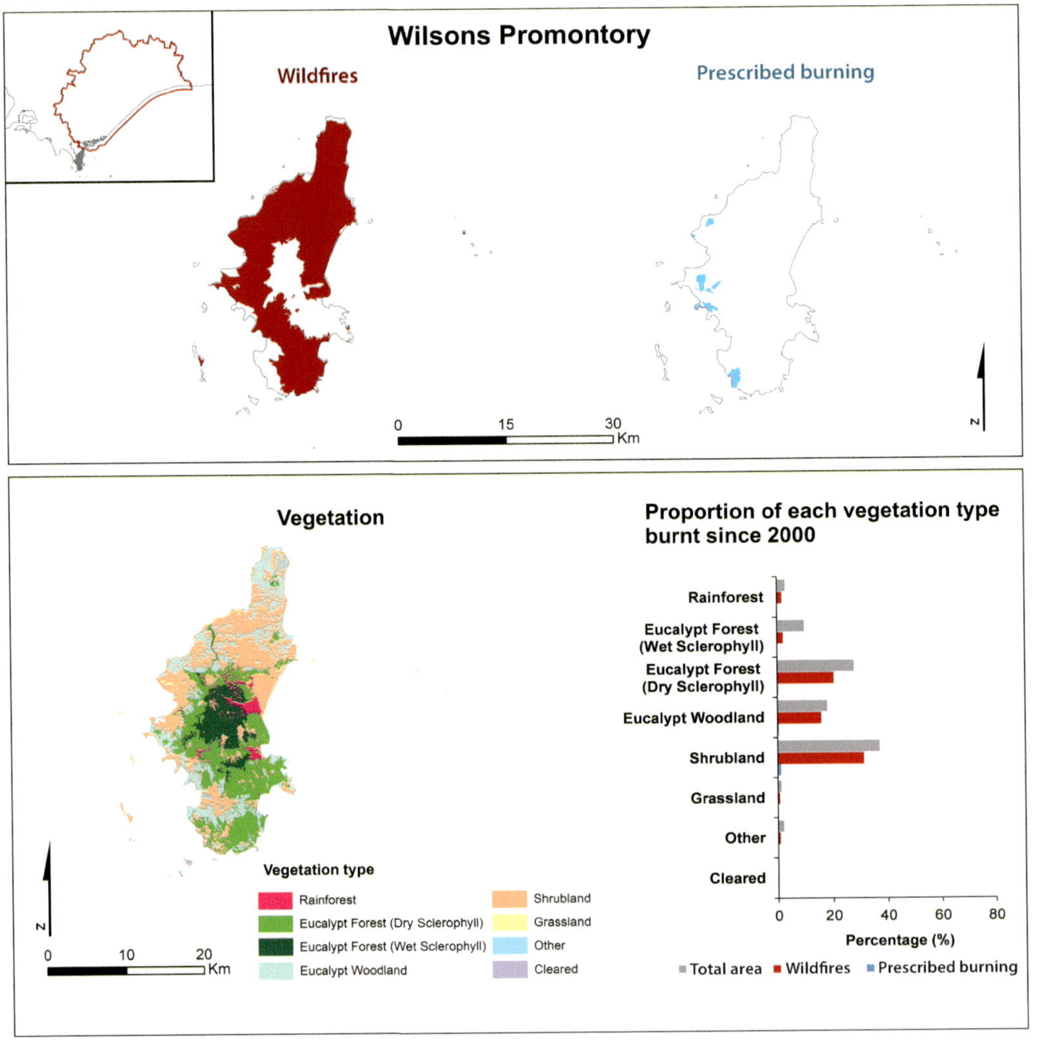

Figure 14.11. Top panel: Extent of wildfires (left, in red) and prescribed burns (right, blue) in the Wilsons Promontory subregion since 2000. Bottom panel: Left: Extent of each vegetation type present in the subregion. Right: Graphic representation of the area burnt since 2000, by vegetation type.

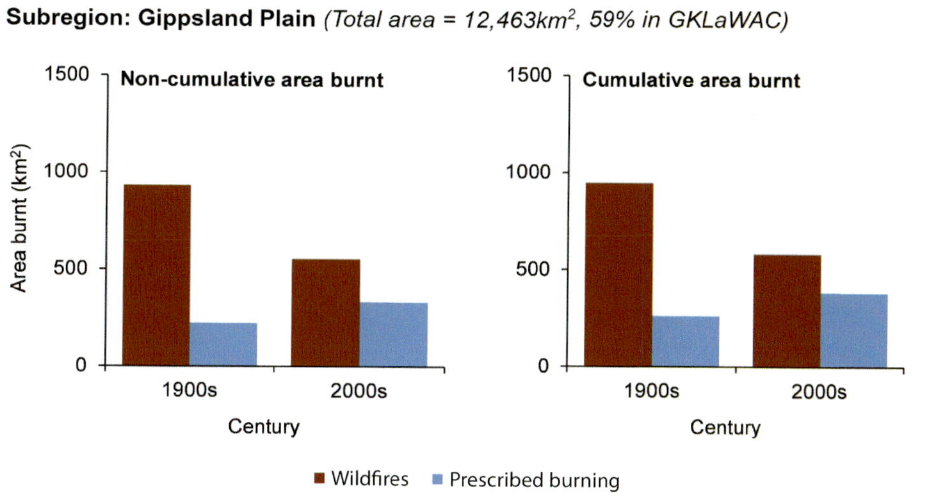

Figure 14.12. Extent of the Gippsland Plain subregion burnt (km²) in the 1900s (1930–1999) and 2000s (2000–2020). Brown bars represent areas burnt by wildfires, blue bars by prescribed burns. The left plot shows total non-cumulative areas burnt during each century (areas may have been burnt more than once), the right plot total areas burnt per year, summed for each century (cumulative); this accounts for areas burnt more than once over those times.

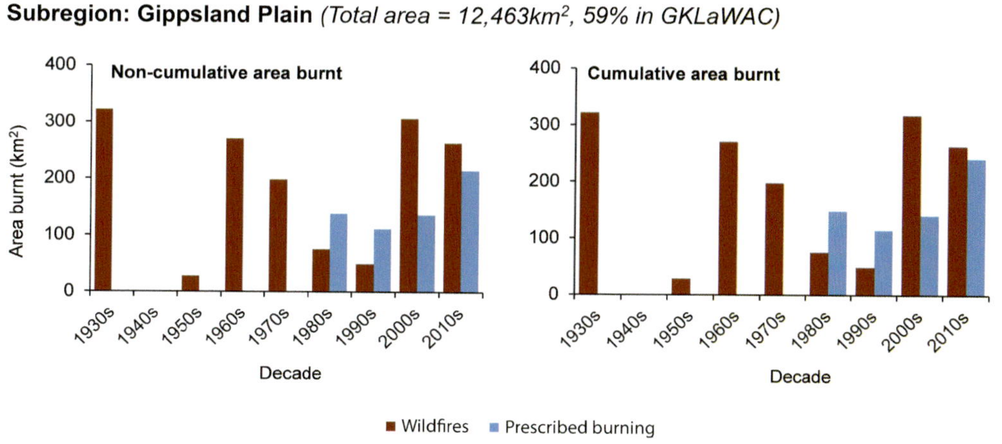

Figure 14.13. Extent of the Gippsland Plain subregion burnt (km²) by decade (1930s to 2010s). Brown bars represent wildfires, blue bars prescribed burns. The left hand-side plot shows the total area burnt in each decade (non-cumulative: areas may have burnt more than once), the right hand-side plot total cumulative areas burnt per year, summed for each decade; this accounts for areas burnt more than once over those periods.

South East Coastal Plain

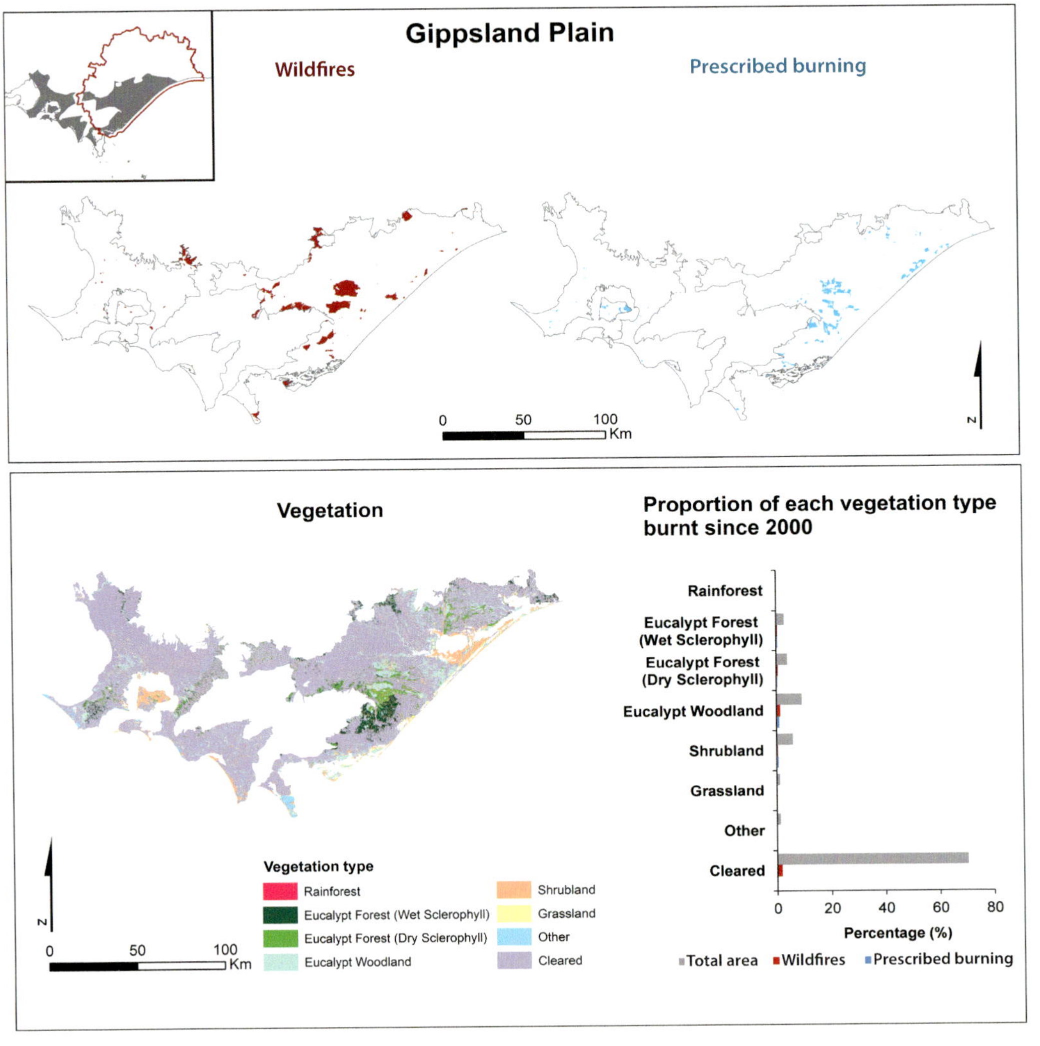

Figure 14.14. Top panel: Extent of wildfires (left, red) and prescribed burns (right, blue) in the Gippsland Plain subregion since 2000. Bottom panel: Left: Extent of each vegetation type present in the subregion. Right: Graphic representation of the area burnt since 2000, by vegetation type.

pattern appears for prescribed burns across the two centuries, again particularly in the 1900s (Figure 14.15).

On average 1261km^2 (20% of the total area) of the East Gippsland Lowlands has been burnt by wildfire in each decade (Figure 14.16). The decade with the greatest area burnt was the 2010s, with 69% (4271km^2) of the total subregion burnt by wildfires—this decade also had notably more cumulative area burnt than any other decade (5327km^2). There was on average 909km^2 of area burnt each decade by prescribed burning, including more than double the area burnt by wildfire in the 1990s and 2000s (Figure 14.16).

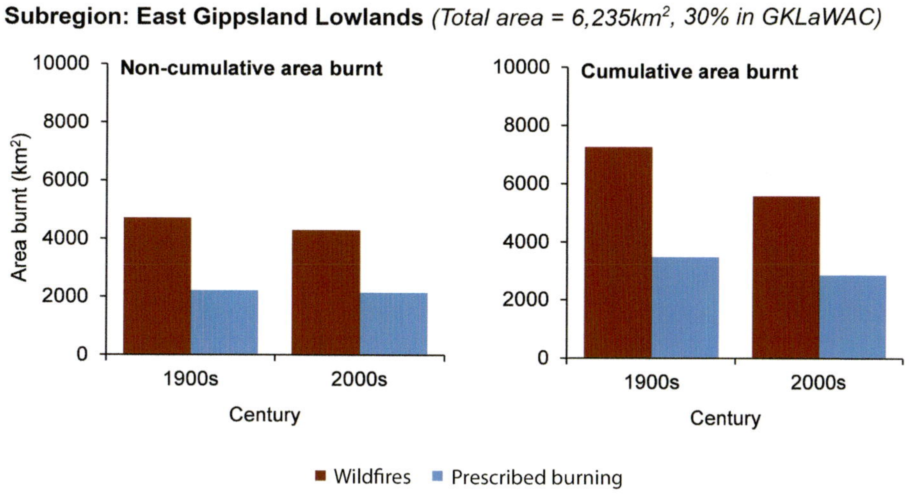

Figure 14.15. Extent of the East Gippsland Lowlands subregion burnt (km^2) in the 1900s (1930–1999) and 2000s (2000–2020). Brown bars represent areas burnt by wildfires, blue bars by prescribed burns. The left hand-side plot shows total non-cumulative areas burnt during each century (areas may have been burnt more than once), the right hand-side plot total areas burnt per year, summed for each century (cumulative); this accounts for areas burnt more than once over those times.

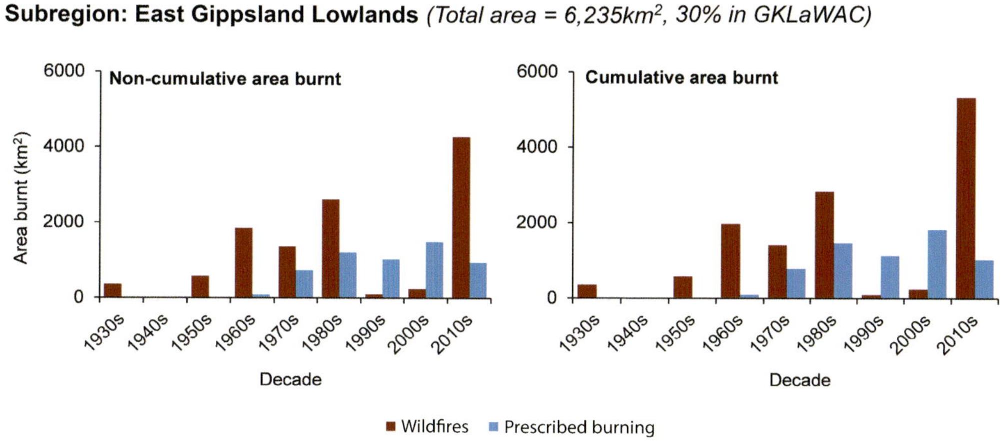

Figure 14.16. Extent of the East Gippsland Lowlands subregion burnt (km^2) in each decade from the 1930s to 2010s. Brown bars represent wildfires, blue bars prescribed burns. The left hand-side plot shows the total area burnt in each decade (non-cumulative: areas may have burnt more than once), the right hand-side plot total cumulative areas burnt per year, summed for each decade; this accounts for areas burnt more than once over those periods.

Since 2000, 69% of the East Gippsland Lowlands has been burnt by wildfires, particularly the eastern half of the subregion (Figure 14.17). Being the dominant vegetation type, the Eucalypt Forest—Wet Sclerophyll had the largest area burnt by wildfires (2222km^2, being 71% of the total area covered by this vegetation type). Almost half (46%) of the total area of Eucalypt Forest—Wet Sclerophyll has also been burnt by prescribed burning (Figure 14.17).

South East Corner

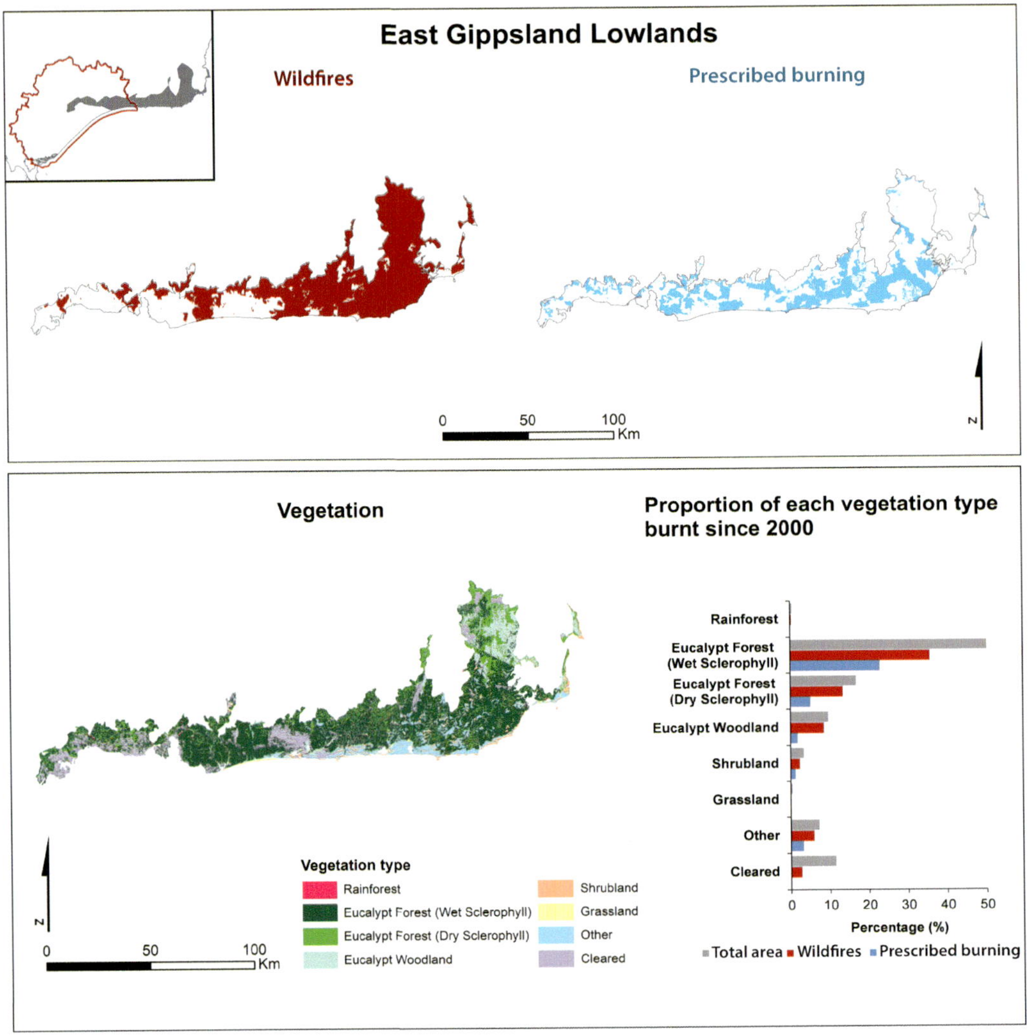

Figure 14.17. Top panel: Extent of wildfires (left, red) and prescribed burns (right, blue) in the East Gippsland Lowlands subregion since 2000. Bottom panel: Left: Extent of each vegetation type present in the subregion. Right: Graphic representation of the area burnt since 2000, by vegetation type.

South East Coastal Ranges. The South East Coastal Ranges subregion has seen more non-cumulative area burnt by wildfires between 2000–2020 than between 1930–1999, but the reverse is true for the area burnt by prescribed burning (substantially more area was burnt by prescribed fires in the 1900s than the 2000s). There have also been many areas burnt more than once by wildfires in the 2000s and by prescribed burning in the 1900s (Figure 14.18).

Wildfires have burnt parts of the South East Coastal Ranges in every decade since the 1930s. The decade with the greatest area burnt was the 2010s (non-cumulative area: 10,803km^2, representing 62% of the total subregion by area), with many areas in this decade also burning

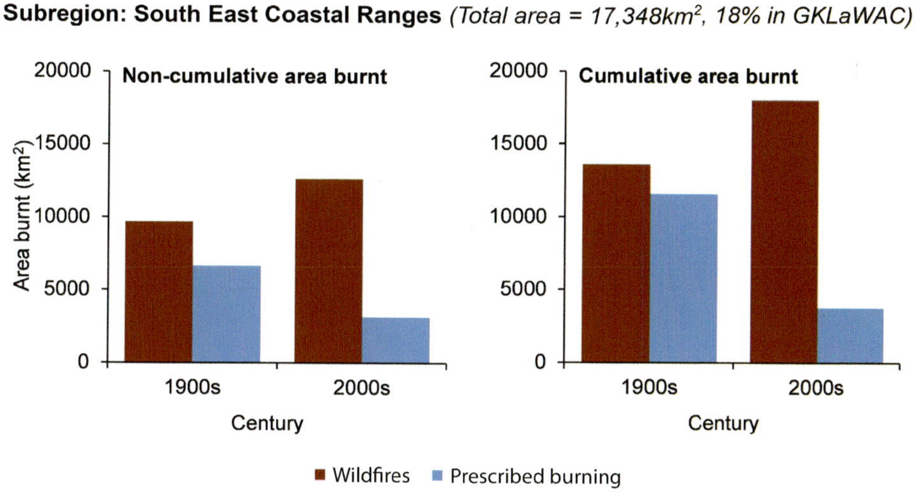

Figure 14.18. Extent of the South East Coastal Ranges subregion burnt (km^2) in the 1900s (1930–1999) and 2000s (2000–2020). Brown bars represent areas burnt by wildfires, blue bars by prescribed burns. The left hand-side plot shows total non-cumulative areas burnt during each century (areas may have been burnt more than once), the right hand-side plot total areas burnt per year, summed for each century (cumulative); this accounts for areas burnt more than once over those times.

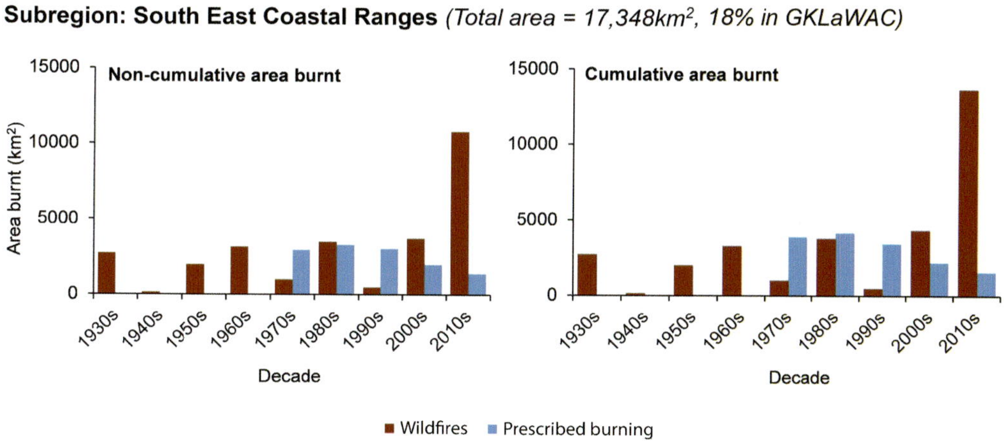

Figure 14.19. Extent of the South East Coastal Ranges subregion burnt (km^2) in each decade from the 1930s to 2010s. Brown bars represent wildfires, blue bars prescribed burns. The left hand-side plot shows the total area burnt in each decade (non-cumulative: areas may have burnt more than once), the right hand-side plot total cumulative areas burnt per year, summed for each decade; this accounts for areas burnt more than once over those periods.

more than once (cumulative area: 13,693km^2). The 1970s, 1980s and 1990s saw greater cumulative areas burnt by prescribed burning than by wildfires (Figure 14.19).

Of the area covered by Cleared vegetation (2354km^2, representing 14% of the total subregion by area), relatively little has been burnt by wildfires since 2000 (472km^2, 20% of the total Cleared land

South East Corner

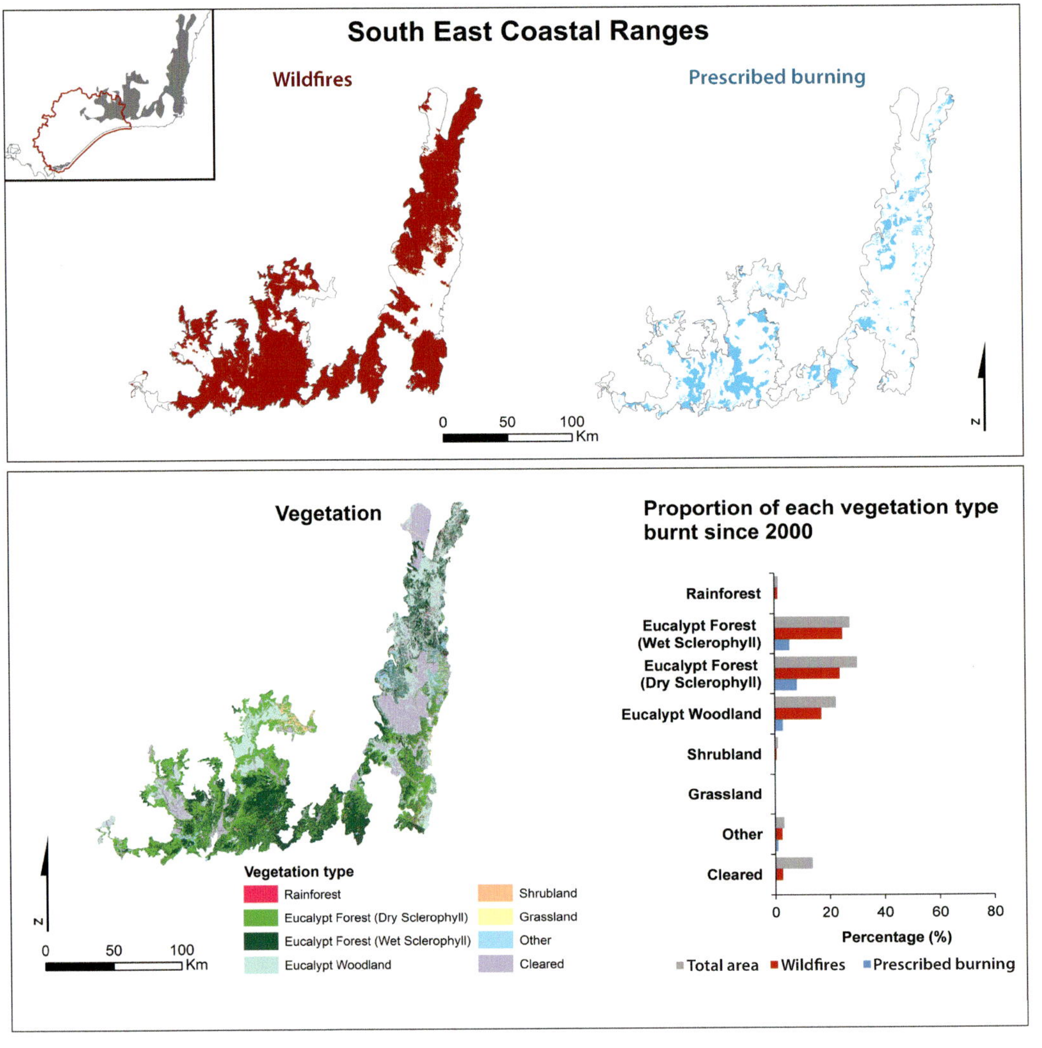

Figure 14.20. Top panel: Extent of wildfires (left, red) and prescribed burns (right, blue) in the South East Coastal Ranges subregion since 2000. Bottom panel: Left: Extent of each vegetation type present in the subregion. Right: Graphic representation of the area burnt since 2000, by vegetation type.

within the subregion). By contrast, the other vegetation types (Eucalypt Forest—Wet Sclerophyll, Eucalypt Forest—Dry Sclerophyll, Eucalypt Woodland) have had on average 82% of their total areas burnt by wildfires. Prescribed burning has mostly occurred in Eucalypt Forest—Dry Sclerophyll, followed by Eucalypt Forest—Wet Sclerophyll, and Eucalypt Woodland (Figure 14.20).

Highlands—Northern Fall. The total area burnt at least once by wildfires is similar for both centuries (non-cumulative 1930–1999: 10,017km^2; 2000–2020: 9260km^2). This is also somewhat the case for areas burnt multiple times (2000–2020 cumulative area: 12,363km^2; 1930–1999: 11,253km^2). However, a noticeably greater area was burnt through prescribed burning in 1930–

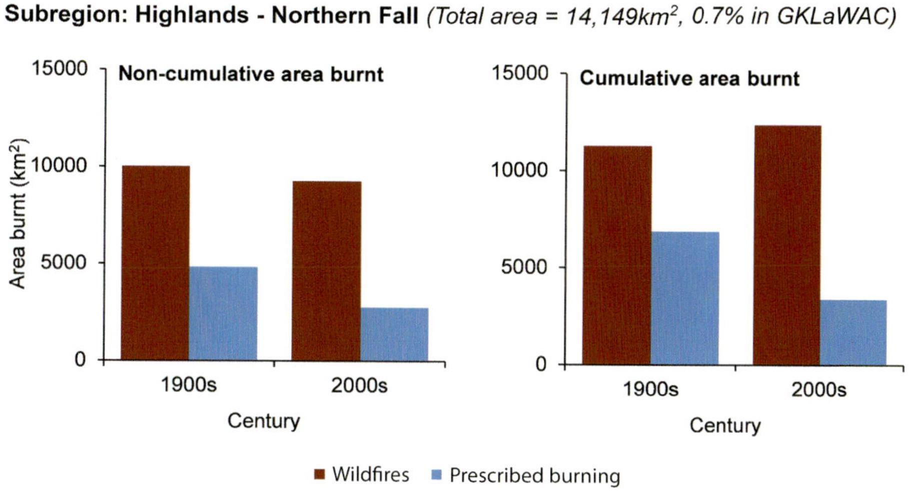

Figure 14.21. Extent of the Highlands—Northern Fall subregion burnt (km²) in the 1900s (1930–1999) and 2000s (2000–2020). Brown bars represent areas burnt by wildfires, blue bars by prescribed burns. The left hand-side plot shows total non-cumulative areas burnt during each century (areas may have been burnt more than once), the right hand-side plot total areas burnt per year, summed for each century (cumulative); this accounts for areas burnt more than once over those times.

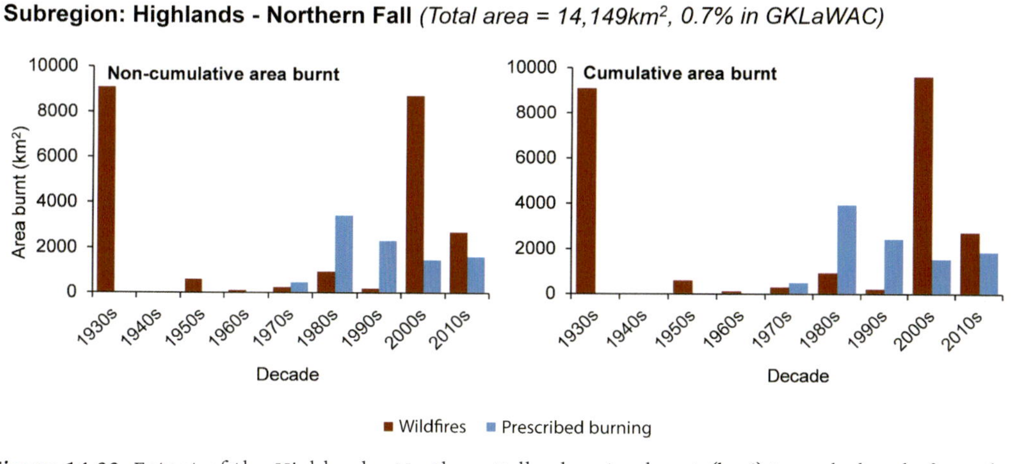

Figure 14.22. Extent of the Highlands—Northern Fall subregion burnt (km²) in each decade from the 1930s to 2010s. Brown bars represent wildfires, blue bars prescribed burns. The left hand-side plot shows the total area burnt in each decade (non-cumulative: areas may have burnt more than once), the right hand-side plot total cumulative areas burnt per year, summed for each decade; this accounts for areas burnt more than once over those periods.

1999 (non-cumulative: 4815km²; cumulative: 6855km²) than in 2000–2020 (non-cumulative: 2740km²; cumulative: 3371km²) (Figure 14.21).

The decade that saw the largest area burnt by wildfires was the 1930s (9095km², representing 64% of the total subregion area), closely followed by the 2000s (8745km², 62%). However, the 2000s was the only decade to have a noticeable increase in the cumulative area burnt

South Eastern Highlands

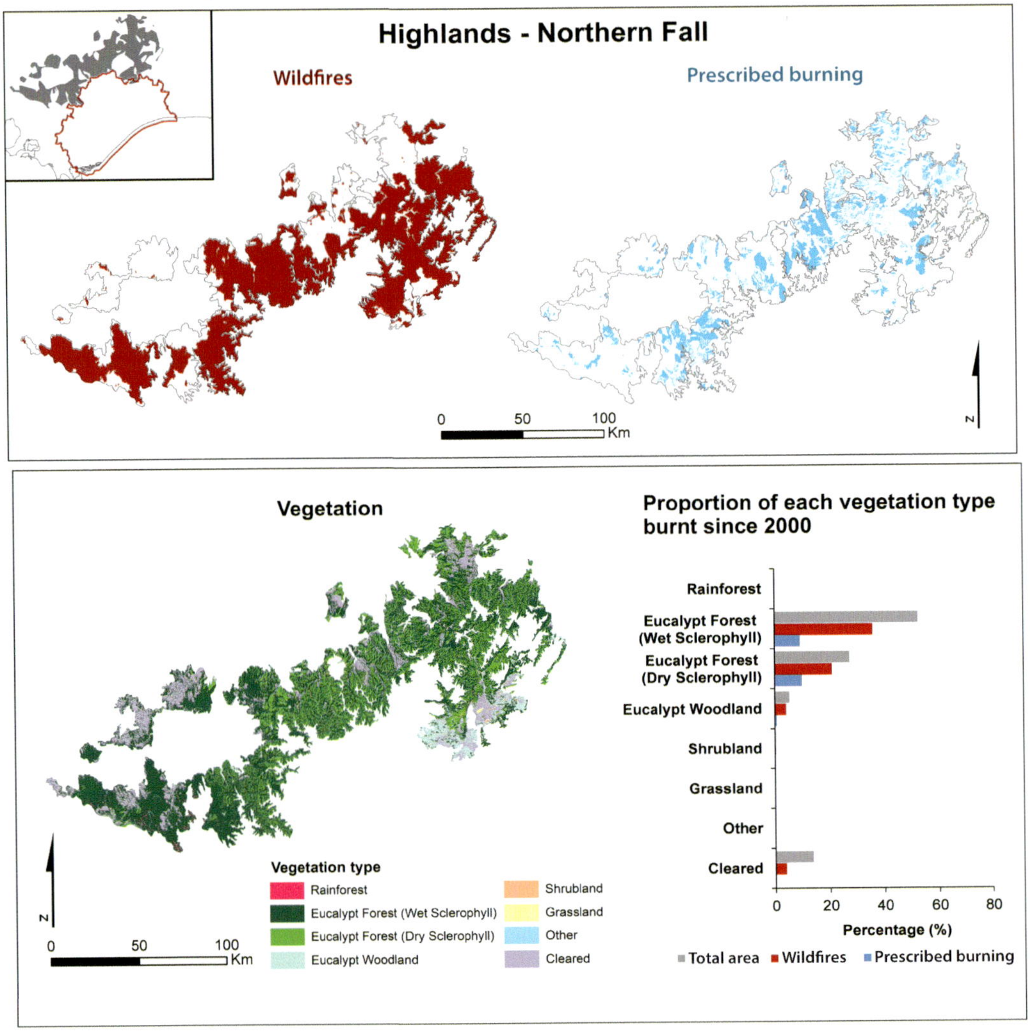

Figure 14.23. Top panel: Extent of wildfires (left, red) and prescribed burns (right, blue) in the Highlands—Northern Fall subregion since 2000. Bottom panel: Left: Extent of each vegetation type present in the subregion. Right: Graphic representation of the area burnt since 2000, by vegetation type.

(9628km²), that is, with an increase in area burnt multiple times by wildfire (Figure 14.22). There were three decades (1970s, 1980s and 1990s) that had greater areas burnt by prescribed burning than by wildfires, and of these the 1980s had by far the greatest, with 3420km² burnt in prescribed events compared to 929km² by wildfires.

Wildfires have burnt 65% of the total area of the Highlands—Northern Fall subregion since the year 2000. Much of this area burnt has been in Eucalypt Forest—Wet Sclerophyll (5082km², representing 68% of this vegetation type), which is the most common vegetation type (covering 52% of the total subregion area), followed by Eucalypt Forest—Dry Sclerophyll (area burnt: 2958km², 76% of the Eucalypt Forest—Dry Sclerophyll area) and Eucalypt Woodland (area burnt: 581km², 77% of the Eucalypt Woodland area). Similar total areas have been burnt by prescribed burning in Wet Sclerophyll and Dry Sclerophyll Eucalypt Forests, despite the higher proportion of Eucalypt Forest—Wet Sclerophyll in the total subregion (Figure 14.23).

Highlands—Southern Fall. The Highlands—Southern Fall subregion has seen a much greater area burnt by wildfires in the 1900s (non-cumulative: 9773km², 82% of the total subregion area) than in the 2000s (non-cumulative: 6928km², 58% of the total subregion area). This pattern also holds true for areas burnt by prescribed burning (Figure 14.24).

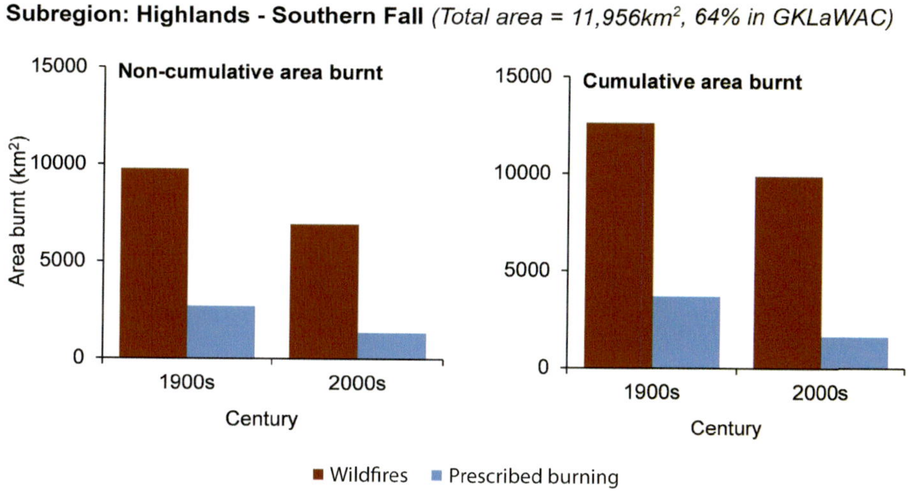

Figure 14.24. Extent of the Highlands—Southern Fall subregion burnt (km²) in the 1900s (1930–1999) and 2000s (2000–2020). Brown bars represent areas burnt by wildfires, blue bars by prescribed burns. The left hand-side plot shows total non-cumulative areas burnt during each century (areas may have been burnt more than once), the right hand-side plot total areas burnt per year, summed for each century (cumulative); this accounts for areas burnt more than once over those times.

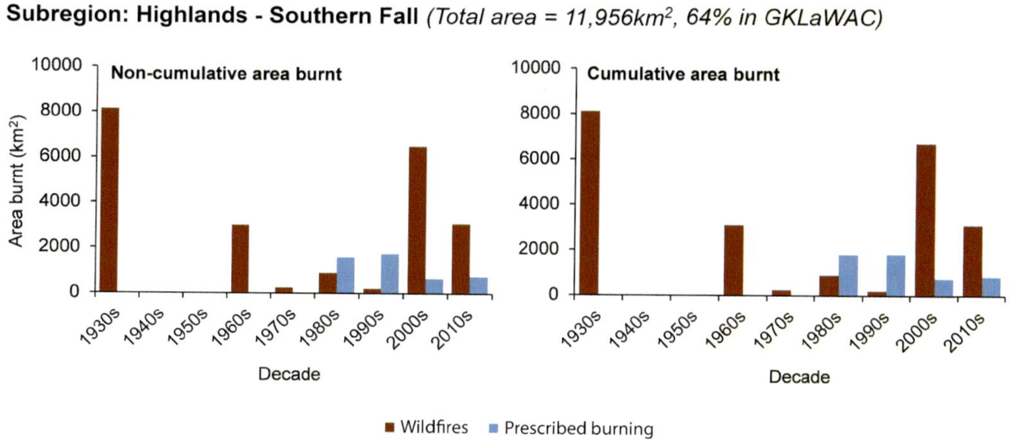

Figure 14.25. Extent of the Highlands—Southern Fall subregion burnt (km²) in each decade from the 1930s to 2010s. Brown bars represent wildfires, blue bars prescribed burns. The plot on the left shows the total area burnt in each decade (non-cumulative: areas may have burnt more than once), the plot on the right the total cumulative areas burnt per year, summed for each decade; this accounts for areas burnt more than once over those periods.

South Eastern Highlands

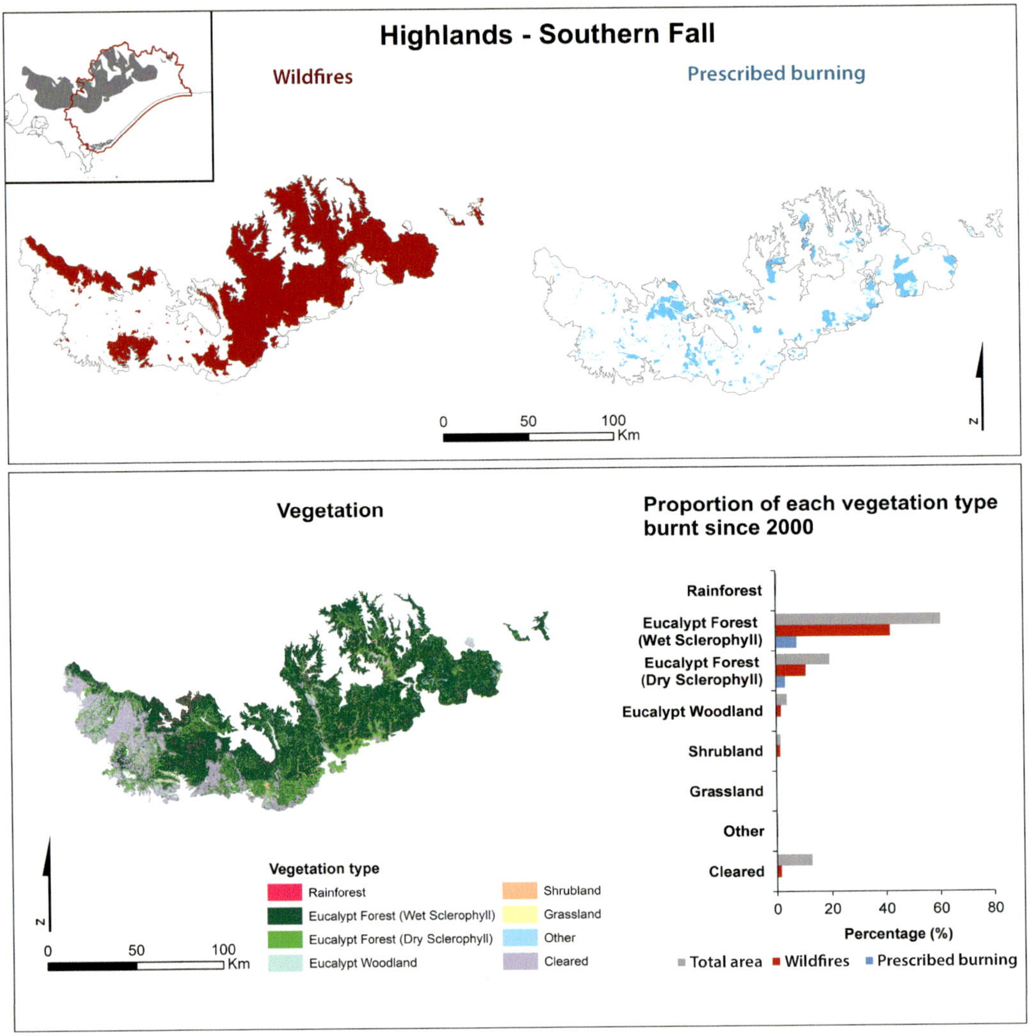

Figure 14.26. Top panel: Extent of wildfires (left, red) and prescribed burns (right, blue) in the Highlands—Southern Fall subregion since 2000. Bottom panel: Left: Extent of each vegetation type present in the subregion. Right: Graphic representation of the area burnt since 2000, by vegetation type.

The decade with the greatest area burnt was the 1930s (non-cumulative area: 8118km², 68% of the total subregion area), followed by the 2000s (6507km², 54% of the total subregion area). There was very little difference in cumulative area burnt during either of these two decades; wildfires did not generally burn more than once in the same locations in these years. The 1940s and 1950s have no record of areas burnt by wildfire. There were greater areas burnt by prescribed fires in the 1980s and 1990s than by wildfire (Figure 14.25).

Over half (58%) the Highlands—Southern Fall subregion has been burnt by wildfire since 2000. As much of the area is covered by Eucalypt Forest—Wet Sclerophyll (61% of the total subregion area), this vegetation type has the highest percentage of area burnt by wildfire (5009km^2, 42% of the total area burnt by wildfire). The western half of the subregion has had fewer wildfires recorded compared to the eastern half, and much of this western area is Cleared land for human use (Figure 14.26). Prescribed burning has occurred mostly in the dominant vegetation types, being Eucalypt Forest—Wet Sclerophyll (area burnt: 917km^2) and Eucalypt Forest—Dry Sclerophyll (area burnt: 404km^2).

Strzelecki Ranges. There has been a slightly greater area of the Strzelecki Ranges burnt by wildfires in between 1930–1999 than between 2000–2020. Nevertheless, the overall area burnt by wildfire in either time period is small (average non-cumulative area burnt: 296km^2, 8.7% of the total subregion area). The area burnt by prescribed burning is even smaller across both periods (average non-cumulative area: 11km^2, 0.3% of the total subregion area) (Figure 14.27).

The only decades that saw prescribed burning were the 1980s, 2000s and 2010s (Figure 14.28). The decade with the greatest area burnt by wildfires was the 1930s, with 318km^2 burnt, representing 9.3% of the total subregion area.

There are only two key vegetation types present in the Strzelecki Ranges subregion, Eucalypt Forest—Wet Sclerophyll and Cleared land (Figure 14.29). The area of Cleared land covers 69% of the total subregion. There is nevertheless very little difference between total areas burnt by wildfires in Cleared land (145km^2) versus Eucalypt Forest—Wet Sclerophyll (123km^2) since the year 2000 (keeping in mind the small proportion of this subregion burnt since the year 2000).

Kybeyan Gourock. The Kybeyan Gourock subregion has witnessed a greater area of burning through wildfires in the past 21 years (2000–2020) than in the previous 70 years (1930–1999) (Figure 14.30). In contrast, a much greater area underwent prescribed burning in the 20th century (729km^2) than in the 21st (174km^2), especially from the 1970s to 1990s. The 1970s to 1990s also saw many locations burnt more than once by prescribed burning (cumulative area prescribed burnt: 1228km^2, 20% of the total subregion area).

The decade that saw the largest area burnt by wildfires was the 2010s (1923km^2, 40% of the total subregion area), followed by the 1980s (521km^2, 11% of the total subregion area). All other decades saw under 500km^2 of non-cumulative and cumulative areas burnt (the 1940s had no registered areas burnt by wildfire) (Figure 14.31). Prescribed burning occurred in every decade since the 1970s, with an average of 218km^2 burnt per decade (maximum: 1980s with 442km^2).

Since the year 2000, wildfires have burnt 41% of the total area of Kybeyan Gourock. Much of this area burnt has been in Eucalypt Forest—Wet Sclerophyll (1234km^2, 57% of the Eucalypt Forest—Wet Sclerophyll vegetation zone), which is the most common vegetation type in this subregion (45% of the total subregion area). Of the 1315km^2 of Cleared land, only 186km^2 has burnt by wildfires. Only a small area (174km^2, 3.6% of the total subregion area) has been burnt by prescribed burning since 2000, and 2% of this occurred in Eucalypt Forest—Wet Sclerophyll areas (Figure 14.32).

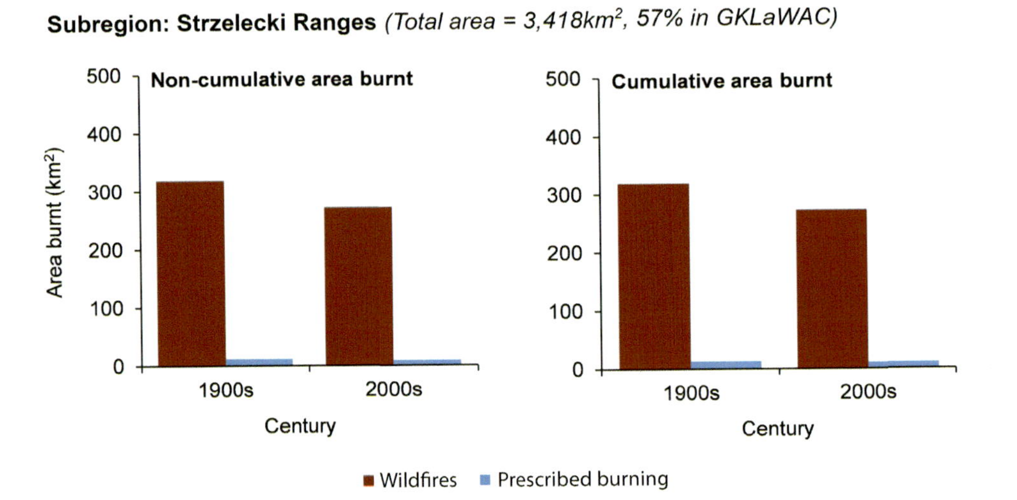

Figure 14.27. Extent of the Strzelecki Ranges subregion burnt (km²) in the 1900s (1930–1999) and 2000s (2000–2020). Brown bars represent areas burnt by wildfires, blue bars by prescribed burns. The plot on the left shows total non-cumulative areas burnt during each century (areas may have been burnt more than once); that on the right shows areas burnt per year, summed for each century (cumulative); this accounts for areas burnt more than once over those times.

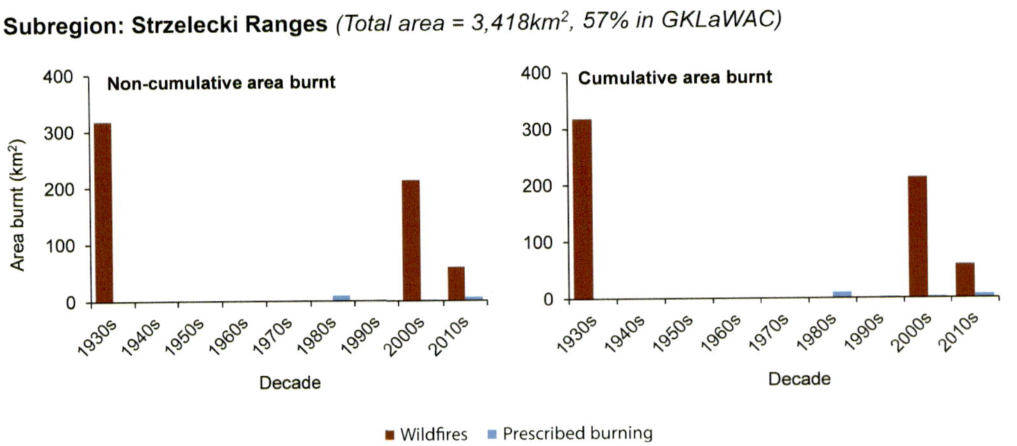

Figure 14.28. Extent of the Strzelecki Ranges subregion burnt (km²) in each decade from the 1930s to 2010s. Brown bars represent wildfires, blue bars prescribed burns. The plot on the left shows the total area burnt in each decade (non-cumulative: areas may have burnt more than once), that on the right the total cumulative areas burnt per year, summed for each decade; this accounts for areas burnt more than once over those periods.

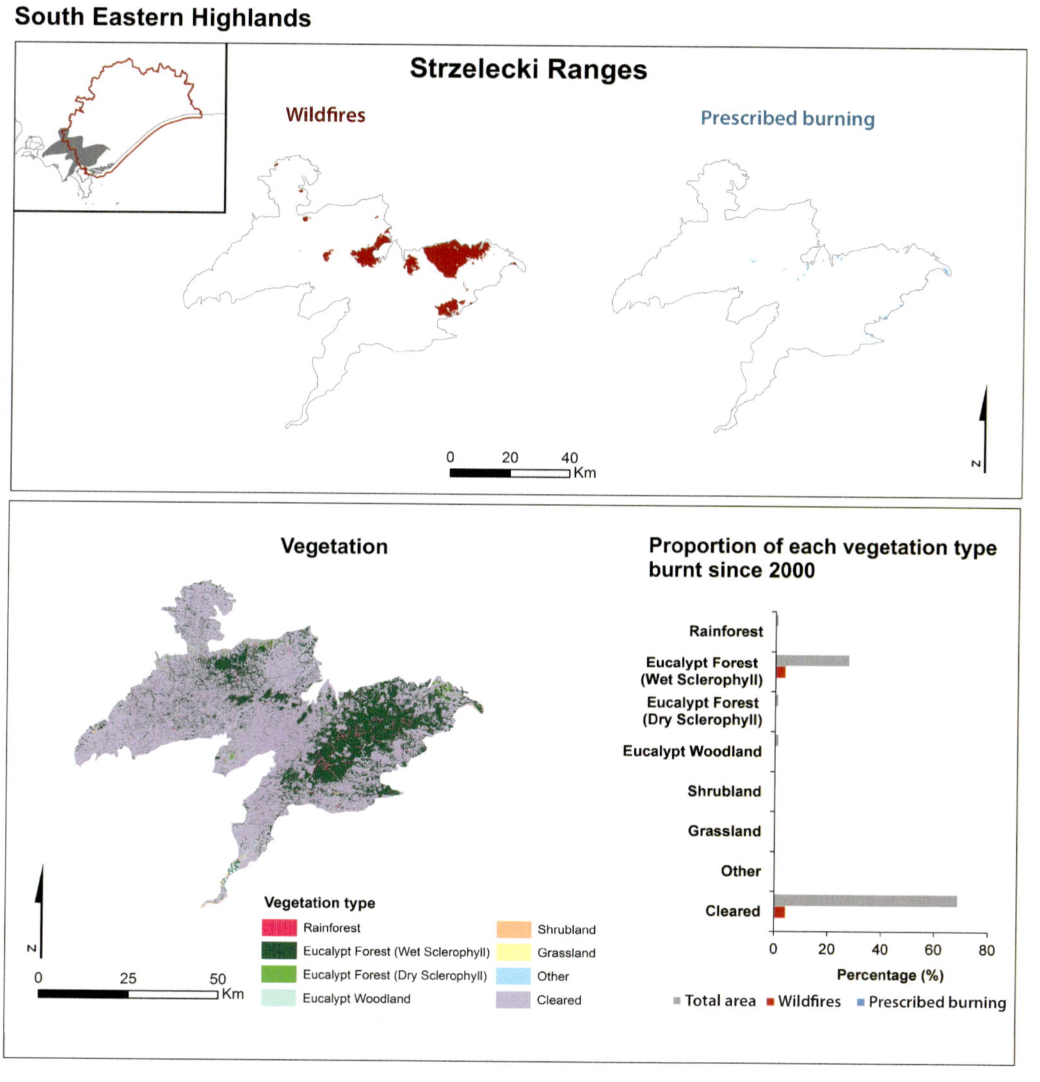

Figure 14.29. Top panel: Extent of wildfires (left, red) and prescribed burns (right, blue) in the Strzelecki Ranges subregion since 2000. Bottom panel: Left: Extent of each vegetation type present in the subregion. Right: Graphic representation of the area burnt since 2000, by vegetation type.

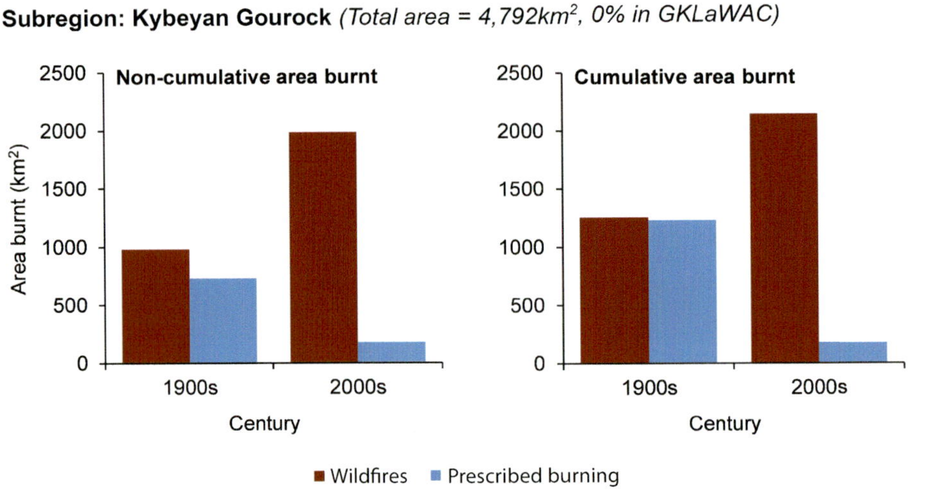

Figure 14.30. Extent of the Kybeyan Gourock subregion burnt (km^2) in the 1900s (1930–1999) and 2000s (2000–2020). Brown bars represent areas burnt by wildfires, blue bars by prescribed burns. The plot on the left shows total non-cumulative areas burnt during each century (areas may have been burnt more than once), that on the right total areas burnt per year, summed for each century (cumulative); this accounts for areas burnt more than once over those times.

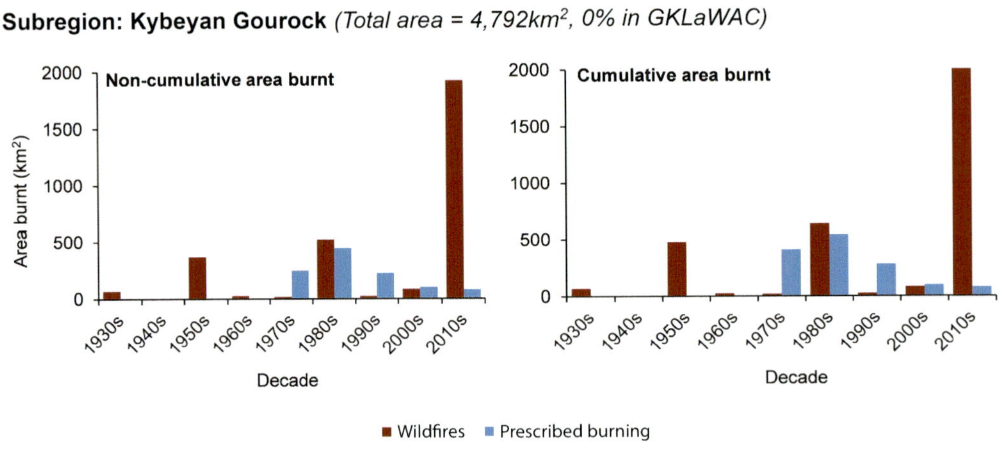

Figure 14.31. Extent of the Kybeyan Gourock subregion burnt (km^2) in each decade from the 1930s to 2010s. Brown bars represent wildfires, blue bars prescribed burns. The plot on the left shows the total area burnt in each decade (non-cumulative: areas may have burnt more than once), that on the right the total cumulative areas burnt per year, summed for each decade; this accounts for areas burnt more than once over those periods.

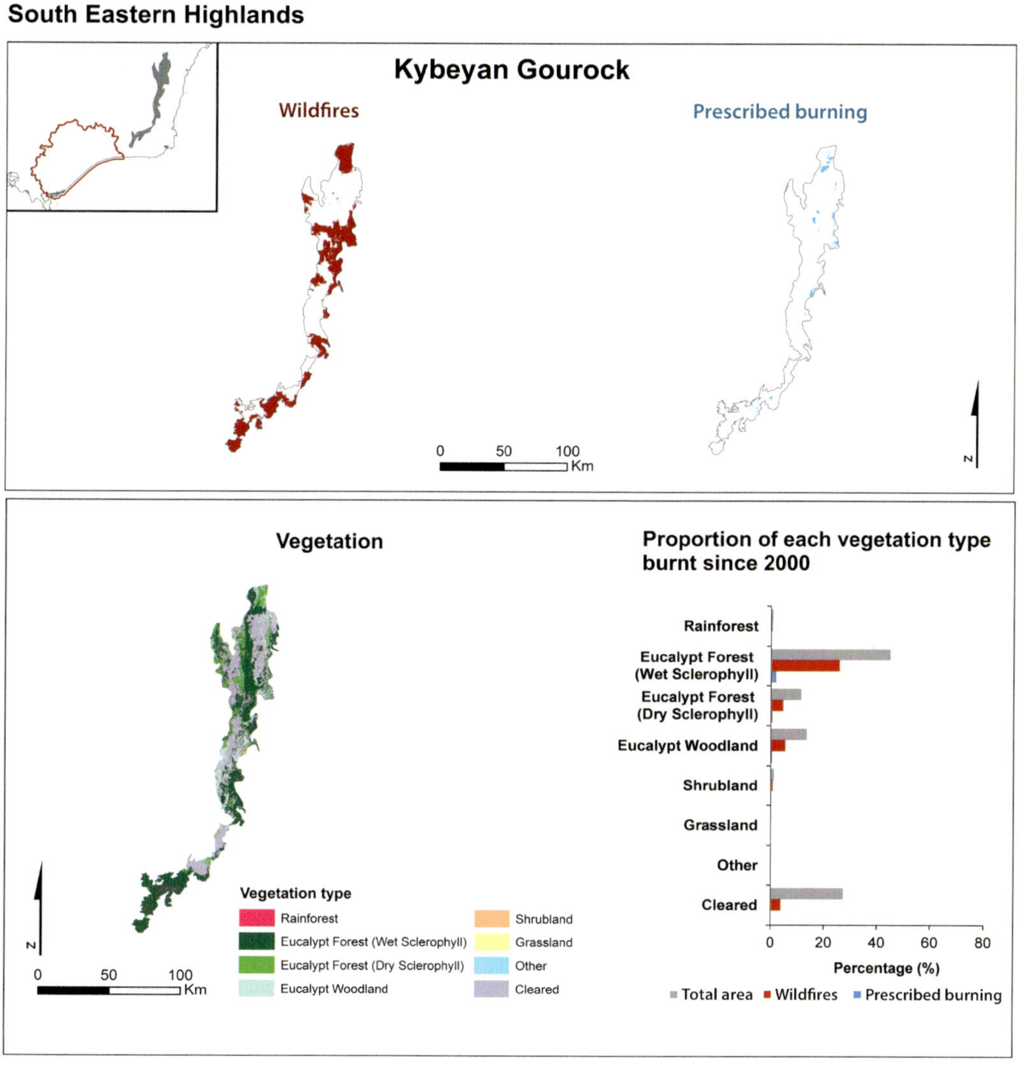

Figure 14.32. Top panel: Extent of wildfires (left, red) and prescribed burns (right, blue) in the Kybeyan Gourock subregion since 2000. Bottom panel: Left: Extent of each vegetation type present in the subregion. Right: Graphic representation of the area burnt since 2000, by vegetation type.

Monaro. Even though about double the area was burnt by wildfires in the 70 years between 1930–1999 as in the past 21 years (2000–2020), a relatively small proportion of the total area of Monaro has burnt during these two periods of time (20% and 10% of the total area, respectively). The areas burnt by prescribed burns over the two timeframes (1900s: 434km²; 2000s: 463km²) are also small, and cover very similar total areas of land, compared to the total size of the subregion. This pattern holds true also for cumulative areas burnt (Figure 14.33).

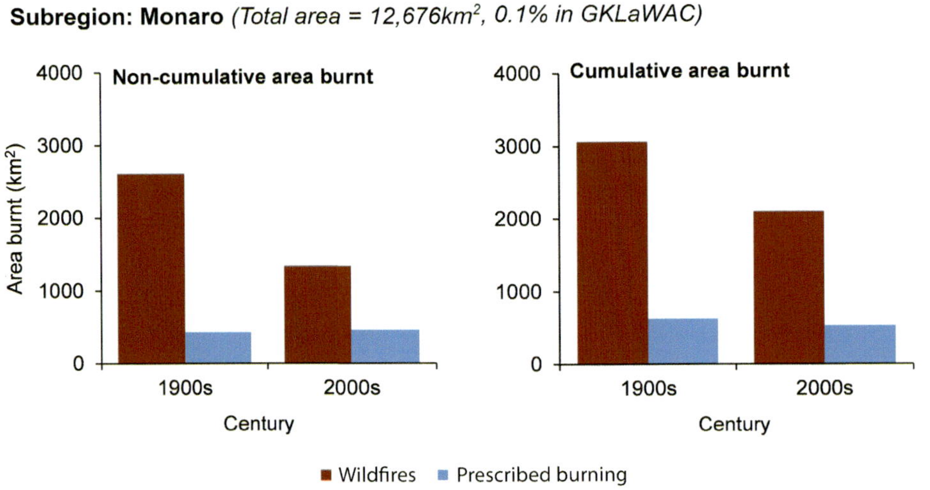

Figure 14.33. Extent of the Monaro subregion burnt (km²) in the 1900s (1930–1999) and 2000s (2000–2020). Brown bars represent areas burnt by wildfires, blue bars by prescribed burns. The plot on the left shows total non-cumulative areas burnt during each century (areas may have been burnt more than once), that on the right the total areas burnt per year, summed for each century (cumulative); this accounts for areas burnt more than once over those times.

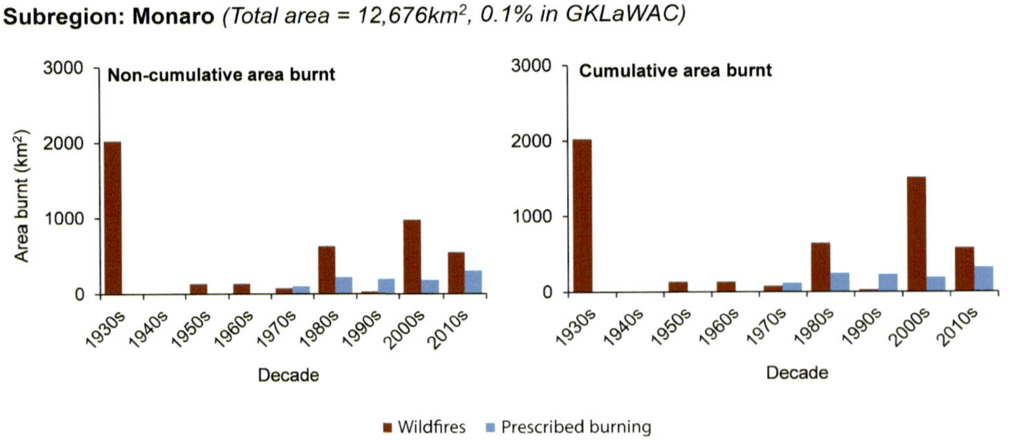

Figure 14.34. Extent of the Monaro subregion burnt (km²) in each decade from the 1930s to 2010s. Brown bars represent wildfires, blue bars prescribed burns. The plot on the left shows the total area burnt in each decade (non-cumulative: areas may have burnt more than once), that on the right the total cumulative areas burnt per year, summed for each decade; this accounts for areas burnt more than once over those periods.

The decade that saw the largest area of land burnt by wildfires was the 1930s (2028km^2, 16% of the total subregion area), the decade of the Black Friday fires. But it was the 2000s that saw the greatest cumulative area burnt (non-cumulative area: 980km^2; cumulative: 1516km^2). Prescribed burning occurred in every decade since the 1970s, with an average of 209km^2 burnt per decade since then (maximum: 2010s, with 312km^2). Both the 1970s and 1990s saw more area burnt by prescribed burning than by wildfires (Figure 14.34).

Wildfires have burnt only 10% of the total area of the Monaro subregion since the year 2000 (Figure 14.35). Of the large area covered by Cleared vegetation (8120km^2, 64% of the total subregion area), relatively little has been burnt by wildfires since 2000 (240km^2, 2.7% of

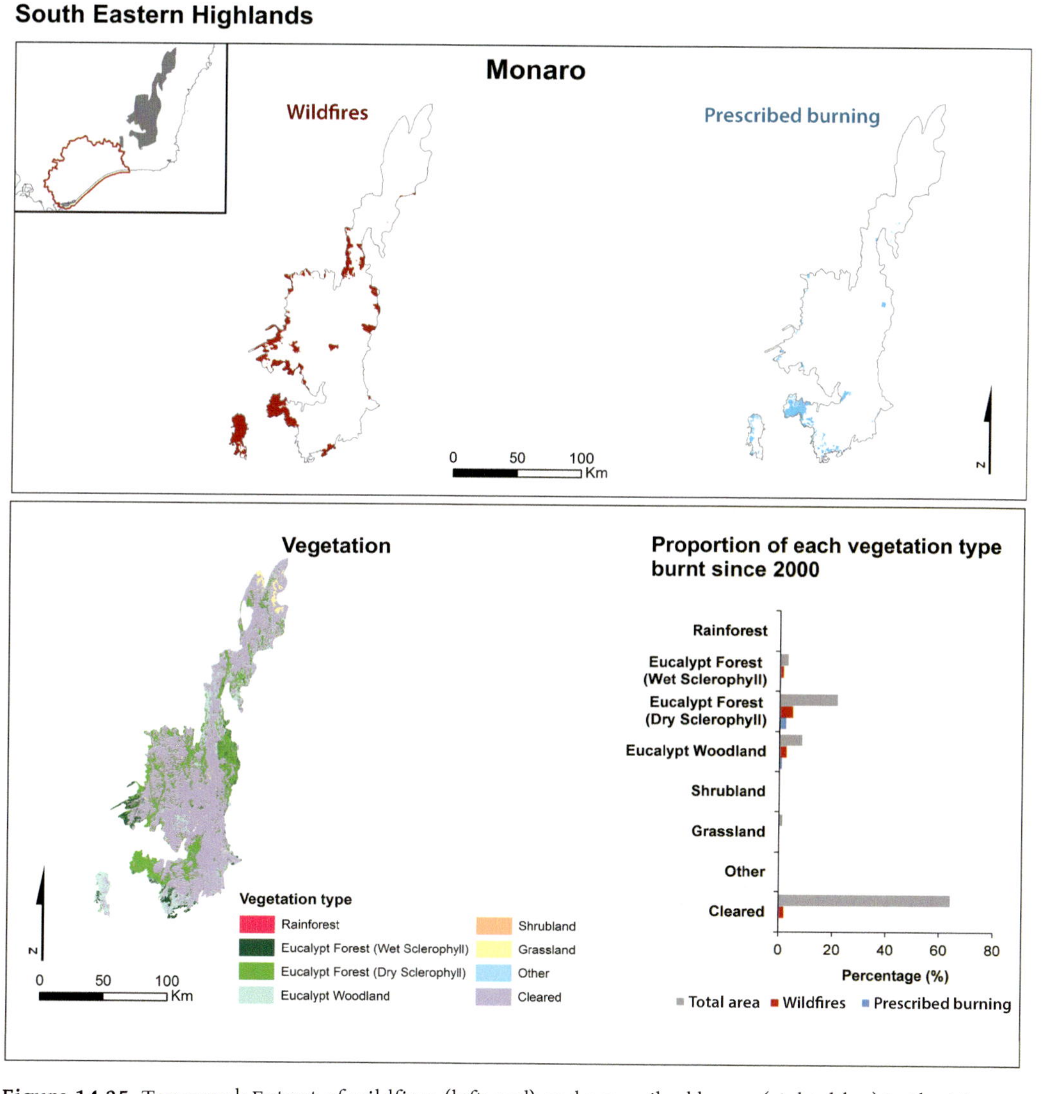

Figure 14.35. Top panel: Extent of wildfires (left, red) and prescribed burns (right, blue) in the Monaro subregion since 2000. Bottom panel: Left: Extent of each vegetation type present in the subregion. Right: Graphic representation of the area burnt since 2000, by vegetation type.

the total Cleared land within the subregion). The Eucalypt Forest—Dry Sclerophyll has had 2.5-times as much area burnt by wildfires (609km²) despite covering only a third of the area (2731km², 22% of the total area of the subregion). Prescribed burning has mostly occurred towards the southern end of the subregion, in the Eucalypt Forest—Dry Sclerophyll and Eucalypt Woodland (Figure 14.35).

DISCUSSION

Across all 11 subregions within or intersecting the GKLaWAC RAP area, we see that the areas burnt by wildfires since the year 2000 is often proportional to the size of the subregion. That is, larger subregions tend to have proportionally larger areas burnt by wildfires (Figure 14.36). However, there are some major exceptions to this pattern: the subregions that have the highest proportions of Cleared vegetation (Gippsland Plain and Monaro) also have substantially smaller areas burnt by wildfire relative to their total areas of land. This again emphasises a point previously observed: that more built-up areas (e.g., settlements, towns, roads, infrastructure, extensive clearing) witnessed less wildfire burning than elsewhere in the GKLaWAC RAP area (see Chapter 5; Figure 5.4). This has implications for the management of GunaiKurnai cultural heritage places, for smaller proportions of the landscape away from areas of development have been surveyed for cultural sites. In other words, the further away an area is from focal points of infrastructure development, the more likely it will be burnt by wildfires, and the less likely its cultural sites will be known, recorded and registered. We address this problem further in Chapter 15.

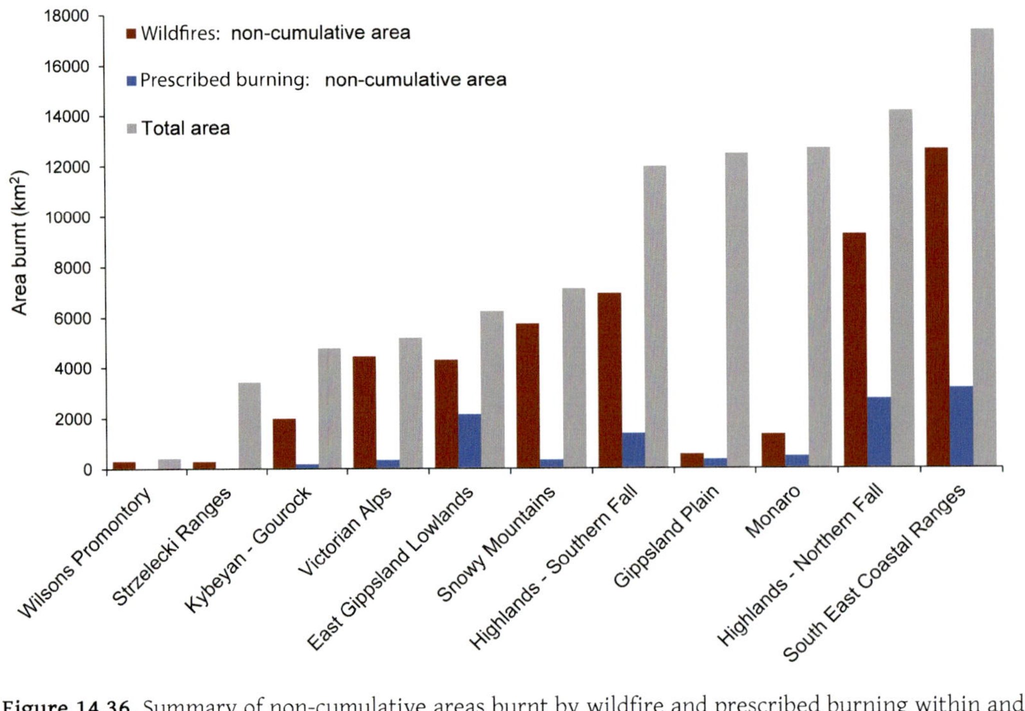

Figure 14.36. Summary of non-cumulative areas burnt by wildfire and prescribed burning within and beyond the GKLaWAC RAP area, by subregion since the year 2000. Brown bars represent wildfires, blue bars prescribed burns, and grey bars the total areas of land covered by each subregion.

Chapter 15

Conclusion

Russell Mullett, Katherine Szabó, Joanna Fresløv, Bruno David, Jessie Buettel, and the GunaiKurnai Land and Waters Aboriginal Corporation

INTRODUCTION

Recent Australian research investigating the potential links between climate change and the scale, intensity and frequency of wildfires is convincingly showing that the frequency of megafire (>1Mha burned) years has increased substantially since 2000, the mean number of years between fires has decreased in each of the past four decades, and that fire seasons are lengthening, with an exponential increase in burned area during the 'cool' seasons of autumn and winter since 1990 (Canadall *et al.* 2021). Research such as this suggests that the increasing numbers of megafires are not a 'blip' but part of a sustained upward trend which is set to continue.

Indeed, the devastating wildfires of the summer of 2019–2020 in Victoria are testimony to this ongoing trend. The span of time over which the wildfires continuously burned was extreme, and their duration led to the fire season being referenced as the 'Black Summer'. East Gippsland in particular was a major centre of intense wildfire activity, meaning that significant areas of GunaiKurnai Country were impacted. Over 60,000 people are thought to have been evacuated from East Gippsland over the course of Black Summer and over 1000 registered cultural sites were impacted (Bushfire Recovery Victoria 2020: 7, 76). Around 13% of the GunaiKurnai RAP area was burned—a devastating loss.

As the scale, severity and duration of wildfires and wildfire seasons are observed to be escalating, it is also becoming increasingly clear that this trend is not restricted to Australia, with wildfire frequency and severity rising in wildfire-prone parts of the world. Wildfires are now emerging as a risk in areas where they were previously rare, and this has become an international problem. The United States is reporting changes analogous to those of Australia, with longer fire seasons and more frequent large wildfires (Westerling *et al.* 2006). An upward trend in wildfire occurrence and severity is predicted to become dramatically worse in Russia (Shvidenko and Schepaschenko 2013), and the decadal frequency of wildfires in the Siberian Arctic has tripled since 2001–2010, with burnt area increased by a factor of 2.6 (Kharuk *et al.* 2022). Surveying global data, the Intergovernmental Panel on Climate Change (IPCC) stated with 'high confidence' that the increasing extent of wildfires was linked to human-induced climate change (IPCC 2022: 45). Moreover, 'for some forest types an increase in the frequency, severity and duration of wildfires and droughts has resulted in abrupt and possibly irreversible changes' (IPCC 2022: 48).

Fires in GunaiKurnai Country (Archaeopress 2023): 191–200

The long history of fire risk, use and management in Australia provides a valuable window onto different management strategies and ideologies, and the various lessons which have been learned over the course of time. The Aboriginal whole-of-landscape approach contrasts with the traditional asset-and-demography focus of European and North American settlers and government—from colonial to contemporary. This in itself has repercussions for the management of heritage, with the official knowledge and recording of heritage sites and places via the Victorian Aboriginal Heritage Register (VAHR) closely bound up with development and the location of populations and infrastructure (see Chapter 6).

FIGHTING FIRE WITH FIRE—DIFFERENT APPROACHES

Outwardly, the designated Victorian state fire authorities—the Country Fire Authority (CFA) covering regional areas, Fire Rescue Victoria (FRV) covering metropolitan zones, and Forest Fire Management Victoria (FFMV) covering state parks, forests and public land—and GKLaWAC share a similar aim of mitigating the regularity and severity of landscape fires. Most notably this includes planned burning to reduce fuel loads, clearing in and around areas of special importance or those required as access-ways, and protecting local environments whilst fostering their renewal through fire (for the latter, see CFA 2011). As Spark drew out in Chapter 3, however, in the past the motivations for and approaches to planned burning by colonial settlers versus GunaiKurnai often differed markedly, and even now certain differences still appear to persist. General public statements such as 'CFA and the Aboriginal community share a culture of fire and have a strong link as custodians of the land and champions of the responsible use and management of fire' (CFA 2011: 5) sometimes mask such differences. Nevertheless, more nuanced observations about Aboriginal approaches to cultural burning, such as preference for single ignition points to slow a fire's build up and allow for the escape of animals (CFA 2011: 6), suggests that a genuine, respectful engagement is developing.

The major authority engaging with the GKLaWAC RAP in the management of cultural burning and wildfire mitigation is FFMV, embedded within the overarching structure and support of the Victorian Department of Environment, Land, Water and Planning (DELWP). Through the Federation of Victorian Traditional Owner Corporations, the landmark 'Victorian Traditional Owner Cultural Fire Strategy' was produced (Victorian Traditional Owner Cultural Fire Knowledge Group 2019). Drawing on diverse local Aboriginal knowledge, it stresses the central importance of cultural burning for Traditional Owners, and articulates a pathway for knowledge-sharing and collaboration with state fire management authorities. The six principles underpinning the strategy (Victorian Traditional Owner Cultural Fire Knowledge Group 2019: 7) align with those that have shone through in this volume:

1. Cultural burning is Right Fire, Right Time, Right Way and for the right (cultural) reasons according to Lore.
2. Burning is a cultural responsibility.
3. Cultural fire is living knowledge.
4. Monitoring, evaluation and research support cultural fire objectives and enable adaptive learning.

5. Country is managed holistically.
6. Cultural fire is healing.

The GKLaWAC approach to cultural burning is outlined in the GunaiKurnai Whole-of-Country Plan (see Chapter 1) and forefronts the same points. The longstanding nature of GunaiKurnai cultural burning and caring for Country have been astutely drawn out by Pullin in her consideration of the early colonial art of Eugene von Guérard: the 'wild' landscapes of colonial settlers were actually the carefully and fully cared-for, managed landscapes of GunaiKurnai Country. This is reinforced by the oral and written histories brought together by Spark in his drawing out of Aboriginal and colonial settler burning and their complex interrelationships (Chapter 3).

KEY HERITAGE VULNERABILITIES

The specific vulnerabilities of heritage places and artefacts to climate change have been considered at international levels (e.g. Hambrecht and Rockman 2017), although to date sea level rise and ice/permafrost melt have received more attention than shifting wildfire regimes. Nevertheless, a long history of studies of the impacts of smaller-scale fires, such as in the heat-treatment of stone raw materials and the processing of food in campfires, provides a useful starting point for considering transformations. Chapter 7 provides technical insights into the means of differentiating fire-affected stone—whether artefactual or unmodified by human hands. With regards to wildfires, everything else being equal buried stone artefacts have a considerably greater chance of intact survival than those exposed on the surface. However, the more sustained heating of higher fuel loads has been shown to have a greater impact on subsurface artefacts (Deal 2012: 110). Larger artefacts are also at greater risk of thermal fracture due to their uneven patterns of heating/cooling (Deal 2012: 110).

Many of the same processes that affect stone artefacts also manifest on rock art and large stone features such as boulders with axe-grinding grooves. These impacts are addressed in Chapter 8 and include thermal shock processes such as spalling and exfoliation. Also as with stone artefacts, fuel load has a direct impact, with larger loads generating more sustained burning and greater negative impacts. 'Cool' burns are thus generally beneficial for rock art sites and, as pointed out in Chapter 8, have the added benefit of inhibiting the growth of lichens and algae. The impact on rock art pigments is variable and depends on the individual properties of the constituent parts. Critical properties include chemical and physical stability, solubility, and the propensity to change colour at high temperatures (for further details, see Chapter 8). The propensity to change colour is well known for ochres, and indeed in some parts of Australia this property is actively manipulated by contemporary Aboriginal peoples as well as by the Old Ancestors. As stated in Chapter 8, this inherent instability causing ochres to change colour when heated (especially from yellow to red) has likely led to the over-representation of red ochres in the archaeological record.

Both bone and shell can be heavily impacted by exposure to fire. Chapter 9 describes the transformational pathways of bone when exposed to different intensities/durations of burning. Complete transformation of bone ('calcination') is rare as ground temperatures during a wildfire rarely reach the temperature threshold of 300 °C, but as with other archaeological materials, heavier fuel loads and the associated sustained burning can indeed cause greater degrees of damage (Chapter 9). Molluscan shell is a more inherently variable material than bone, with some types of shell being highly susceptible to fire damage while others are more resistant (Chapter 11). As outlined in Chapter 11, experimental work has shown that shells with laminate constructions (such as mussels and oysters) are very vulnerable to fire damage and will tend to shear into thin fragments when impacted by high temperatures. Chemical transformation is a standard occurrence, and this has the potential to impact other sorts of analysis such as radiocarbon dating and stable isotope analysis. Burial under sediments does seem to afford some protection from both mechanical and chemical changes.

Scarred trees, by their very nature, are highly susceptible to damage and destruction by wildfires. As outlined in Chapter 10, it is very difficult to know how wildfires have impacted the numbers and distributions of scarred trees in the landscape, due to both a deficit in recording over the past few decades, and the fact that surveys have tended to focus on the accessible lowlands and immediately along access pathways in the High Country rather than across broader areas. Of the 387 trees listed on the VAHR in the GunaiKurnai RAP area, only 30 are known to be currently present in the landscape (Chapter 10). The rarity of large, older, River Red Gums and Stringybark trees—the preferred species for GunaiKurnai bark removal for canoe manufacture—also suggests that the number of remaining scarred trees will be very low.

While there has been no research into the chronology of Scarred Trees in GunaiKurnai Country, it is likely that the surviving Scarred Trees represent a very short and recent period of time in the long history of occupation of East Gippsland, possibly as little as 500 years. As such, they span the time of the first arrival of the colonial explorers and subsequent settlement. They record the travels and activities of the Old People both on land and water before, during and, to a lesser extent, after the early colonial period. Untouched or disturbed, they can last hundreds of years, but they will inevitably succumb to death and destruction. Once damaged by fire, the process of deterioration speeds up from a slow to swift decline. While they are less susceptible to damage during post-fire clean-up activities, as most people in rural areas are familiar with their cultural values to Aboriginal people and can identify scarred trees, they nevertheless remain among the most vulnerable traces of the past. The result of the increasing frequency and severity of fires is a rapid destruction of these cultural icons, so that if the current rate of burning continues, culturally modified trees in GunaiKurnai Country may be reduced to a few protected trees in towns and photographs of what was once there.

KNOWING WHAT WE DON'T KNOW

The unknowns surrounding past and present distributions of Scarred Trees gives focus to a wider problem: the patchiness of archaeological and heritage surveys, and the resultant unevenness in the data. Indeed, there are several levels of issues with the basic distributional data on cultural sites and their contents, including (1) information locked in 'grey' literature; (2) variability in terminology and categories within the VAHR over time; (3) the spatial unevenness of the survey data; and (4) variability in the detail of recording. These are all briefly discussed here.

Survey of the grey literature. Almost all the sites listed on the VAHR were recorded, and registered, as a result of cultural heritage management consultancies or casual observation. Very few were recorded through academic research. One corollary of this is that the contexts of their field 'discovery' are locked away in a voluminous sum of grey (unpublished) literature. This matters when doing the kinds of analyses we have presented in this monograph, because to understand spatial patternings, we should ideally be relating site distributions against where people have looked and systematically recorded what they found or what they knew at the time. We have not been able to do that here, because of the elusiveness of significant numbers of unpublished reports. This problem is compounded by the fact that archaeological surveys in GunaiKurnai Country have varied enormously in quality over the years. Sometimes, a site will have been recorded as a result of careful field walking by closely spaced archaeological surveyors. At other times, a single or team of field surveyors will have driven across the landscape looking out for sites from their car windows, only going out to record a site when they saw something from the comfort of the front seat, driving at slow speeds. Some 'survey reports' are not field surveys at all, but compiled summaries of site distributions based on the sites registered on the VAHR at that time.

Cross-cutting against this variable quality of site surveys, is an equally variable quality of the skills and attentiveness of the practitioners who recorded the sites. In the end, there has been no systematic compilation of precisely where people have gone to record sites, nor of the survey strategies they have used. While we may know from experience and participation where some of the individual surveys were done and what was found, for the kinds of systematic analyses presented in this monograph we do not know where exactly all the surveys took place, how systematic they were, whether they used quadrats or transects, followed systematic random or selective survey areas, whether those survey areas were stratified or not, whether people just looked along walking tracks or covered broader areas and so forth. This does not deny the site distributional patterns reported in this volume, but it does mean that a next step of research needs to methodically plot the exact areas where site surveys have been undertaken, and the nature and quality of those surveys. By doing so, the degree to which the patterns in the results (the registered sites), as presented in this monograph, map onto and can thus be said to have been influenced by landforms, vegetation communities, bioregions and so forth, rather than simply by where people have looked, can be determined. In short, biases created by sampling can be detected, assessed, and their impacts understood.

VAHR terminology and categories. As outlined in Chapter 6, there have been changes to some of the VAHR site types ('components') through time. Most notable, and intractable, is the 2012 addition of 'Low Density Artefact Distributions' (LDADs), where fewer than ten stone artefacts are found within an area of $100m^2$, alongside the categories 'Artefact Scatter' and 'Isolated Artefact'. This has made it impossible to aggregate data from before and after 2012 in any reliable way—a key issue given that Artefact Scatters represent far and away the most common type of archaeological site in the GunaiKurnai landscape (see Table 6.1). In some instances, it may be possible to go back to original records to source the data on the numbers, types and spread of stone artefacts at any given Artefact Scatter or LDAD listed in the VAHR. However, as explained in Chapter 6, for LDAD sites all artefacts must be analysed, whereas for Artefact Scatters only a 'representative sample' is required. This circles us back once more to potential incomparability.

Spatial variability in survey data. The number of registered cultural sites burnt by wildfires is strongly correlated with the total area burnt. That is, the greater the area a wildfire has burnt, the greater the number of known cultural sites it has burnt in the process. This has been a general pattern observed with each large wildfire over the years (e.g., Chapter 5: Figure 5.2). While the total area burnt across the GKLaWAC RAP area over the 21 years since 2000 is slightly less than during the previous 70 years, it has done so at twice the burning rate. Furthermore, nearly half (44%) of the registered cultural sites have been burnt more than once, with some up to five times (Chapter 12). While these patterns signal a worrying trend that show no signs of abating, the variability in the spatial distribution of the underlying survey and spatial data and the potential for future heritage surveys also need to be considered. Hence, for any large-scale spatial analysis, the ability to make strong inferences on observed patterns across the landscape largely depends on the quality, quantity, and spatial and temporal resolution of the underlying data. However, for most cultural heritage surveys, there are challenges with logistics, accessibility to sites, cost, and time constraints that greatly limit (and sometimes prohibit) the amount and quality of data that can be collected. As a result, surveys are often done in areas of easiest access (i.e., closer to denser population centres or infrastructure), and many developments requiring cultural heritage surveys also happen to be near already existing infrastructure (because infrastructure often brings with it further infrastructure needs and ongoing developments that trigger legislated cultural heritage survey requirements) (Figure 5.4). Similarly, as seen in Chapters 6 and 14, where there is a higher human footprint there is also a greater degree of land clearance, more registered cultural sites, and fewer fires (and more focused fire-fighting to limit the extent of fires when they do threaten the area). When it comes to the known, recorded distribution of Aboriginal cultural heritage sites, this creates a circular or compounding bias in the data and subsequent analyses, one that is problematic in respect to the kinds of issues addressed in this monograph, by the fact that major wildfires generally occur in areas away from such populations and infrastructure. This means that the impact of wildfires on heritage-at-large is difficult to quantify with the current site location data at hand:

> **Most of the available data about cultural sites (i.e., the registered cultural sites) are close to population centres and infrastructure, but most of the wildfires occur away from population centres and infrastructure, so that the sites that need most protective action from wildfires are unregistered and largely unknown as a result.**

It is also worth noting that while details of the precise locations of the archaeological surveys undertaken by previous field researchers lie scattered in dozens of unpublished reports and have not been compiled or analysed, Elder Uncle Russell Mullett and archaeologist Joanna Fresløv—both of whom have witnessed most of the cultural heritage surveys undertaken across GunaiKurnai Country over the past few decades—have noticed a clear pattern: most surveys have been undertaken along recent vehicle tracks. This may have biased the survey results (because the tracks often follow ridge-lines or other particular landscape features), the degree to which surface artefacts were burnt (because tracks are often lightly vegetated and have more sparse leaf litter), and the amount of damage cultural heritage sites have been subjected to as a result of post-fire heavy-vehicle activities favouring tracks. Future surveys should thus also include areas away from tracks such as along slopes, river terraces etc., to better determine patterns across the landscape.

Variable record detail and quality. Within the Victorian fire spatial data, there are also issues which stymie further detailed analysis of patterns, including the lack of data for the GunaiKurnai RAP area from 1903–1927, and the fact that only 19 Traditional Owner fires ('cultural burns') are recorded in the Victorian fire spatial data—all dating from 2019 onwards (see Chapter 5 for further detail). However, many of these inconsistencies and issues in the underlying data are unavoidable and largely commonplace among historical data aggregated from several different sources. These fire data for Victoria were primarily developed using collated and digitised historical records (e.g., from record searches, historical aerial photographs, regional fire records), aggregated in a form for use in a Geographic Information System (GIS) for spatial analysis. As is inherent with historical data and the compilation process, there are issues with infrequent data reporting and past record-keeping that make standardisation and digitisation difficult. The error range documented for the Victorian fire spatial data is 10–300m, with such errors compounding when overlapped with other spatial data (e.g., vegetation type and land-use: Chapters 6 and 14) where there can be mismatches between the scales of each spatial layer that can be difficult to identify or rectify. This means that interpretations of patterns using spatial data are rarely exact, giving instead approximate indications of patterns in the spread of fires, total areas burnt and overlaps with landscape characteristics. So, while these data are the most comprehensive that can be obtained on historical wildfire and prescribed burning events, it should be acknowledged that there are inherent caveats (e.g., re-scaling of spatial layers during compilation). Additionally, while Victoria's subregions give us great ecological context of the landscape in southeast Victoria, there is only so much we can infer from their underlying ecology (represented as spatial layers of vegetation type and land-use) to fire patterns at such broad spatial scales. This is partly because of the inaccuracies in the development and projection of spatial layers but also due to the environmental heterogeneity that exists within each subregion at finer spatial scales. Addressing these limitations and issues is particularly important for the more fire-sensitive vegetation types such as temperate rainforests (Chapter 14)—which typically occur in very small patches within GunaiKurnai Country and broader southeastern Australia—where accurately identifying fire history and risk is critical to the management and protection of these landscapes and the cultural sites they contain.

There are fewer records of cultural sites in areas where the human footprint is low (e.g., Figure 5.4), a pattern that is also reflected elsewhere across southeastern Australia. Of the seven biogeographic subregions that have large areas that intersect with the GKLaWAC RAP area (the other four subregions are either just within or just outside the RAP area), there is thus a need to focus future heritage surveys in areas where cultural sites are under-represented on the register. This would clarify whether sites in these areas are truly under-represented by their low densities on the VAHR as a result of limited past surveys in those areas. These subregions of low site representation warranting dedicated cultural heritage surveys are: Highlands—Southern Fall, South East Coastal Ranges, Strzelecki Ranges and the Victorian Alps (Figure 14.2). More specifically relating to habitat, very few cultural sites have been registered from Eucalypt Forest—Wet Sclerophyll (Figure 6.8) and from areas of low-intensity human use (Figure 6.9).

With regards to archaeology, our understandings of the impacts of fire on archaeological materials have historically been informed by research questions around small-scale burning such as the heat-treatment of stone raw materials (see Chapter 7 and references therein) and the use of fire in food processing (see Chapter 9 and references therein). Experimental and fine-grained studies provide valuable starting points in understanding the response of heritage materials and assemblages to larger-scale burning, but landscape-level fires present an additional series of risks and challenges not captured in smaller-scale studies. Most pertinently, these additional concerns include the destabilisation and heightened erosion of surface sediments following a wildfire, exposing archaeological materials to increased weathering and risk of movement (see Chapter 2). As part of wider firefighting and wildfire mitigation practices, the impacts of heavy machinery involved in processes such as the making or maintaining of fire-breaks and turn-around areas on ridge-line access tracks, also have well-recognised and demonstrable impacts on Aboriginal cultural heritage sites (Chapter 2; see also Timmons *et al.* 2012).

LOOKING TO THE FUTURE

As the impacts of climate change are ever more clearly felt in day-to-day life, the need to actively predict risk, reassess action plans and move to protect that which we value becomes more pressing. Different countries, governments and organisations have responded in a variety of ways, and with variable cohesion and urgency. In terms of policy and coordinated action regarding the protection of heritage in the face of climate change, in some ways the United States has visibly been on the front foot. In 2014, the Union of Concerned Scientists released 'National Landmarks at Risk: How Rising Seas, Floods, and Wildfires are Threatening the United States' Most Cherished Historic Sites' (Holtz *et al.* 2014), which acted as a call to arms with a selection of case studies focussed on risks to well-known heritage sites. However, even before this the protection of heritage from wildfires had been proactively tackled by the U.S. Department of Agriculture, with initiatives like their 'Wildland Fire in Ecosystems: Effects of Fire on Cultural Resources and Archaeology' (Ryan, Trinkle Jones *et al.* 2012) being a touchstone. The U.S. National Parks Service has also sought to mobilise the latest scientific understandings regarding climate change and activate them in policy and practice, with a policy memorandum, 'Climate Change and Stewardship of Cultural Resources' (U.S. National Parks Service 2014) setting a direction which has now led to the fully developed 'Cultural Resources Climate Change Strategy' (Rockman *et al.* 2016). Amongst other proactive engagements in the U.S. was the Society for American Archaeology's formation of the Climate Change Strategies and Archaeological Resources Committee in 2016—an initiative which could usefully be mirrored in Australia.

While many of the issues faced in the U.S. are analogous to those faced in Australia, such as the spatial scale of landscapes and prevalence of wildfires in particular locations, other issues are distinct. Coniferous forests and layers of 'duff' (dried conifer needles) on the ground stand in sharp contrast to Victoria's rainforests and fire-adapted eucalypt forests and woodlands. The cultural context of landscape burning also has to be central to thinking through wildfire management, and indeed landscape management in the most general sense, given the preeminent importance of burning to life, lore and well-being among GunaiKurnai and other Traditional Owners across Victoria. In essence, the whole GunaiKurnai worldview contrasts with conventional governmental structures and approaches to management. As discussed in this volume, and clearly stated in the 'Victorian Traditional Owner Cultural Fire Strategy'

(Victorian Traditional Owner Cultural Fire Knowledge Group 2019), the GunaiKurnai and Victorian Traditional Owner perspective is holistic—the landscape and its peoples, places, plants and animals are not divisible into parts to be considered and managed separately. This is as relevant for wildfire management as it is for heritage protection and management, with the VAHR presenting nodes as isolated heritage sites across the landscape, whereas GunaiKurnai see the entire landscape as an interconnected cultural whole. Moreover, both the parts and the totality come with a duty of care.

GunaiKurnai bring the experience and knowledge of generations to managing severe wildfire risk and using managed fire to support and care for people, Country, and its embedded Ancestors. However, the forced dislocations from Country, policies of suppression of burning, and landscape transformation of some two centuries of colonial jurisdiction, farming and land management, present a different contemporary Victorian reality. The certainty of human-induced change is also rapidly shifting the baselines by shortening the season for 'cool' burns as the risk of megafires rises. The 'Victorian Traditional Owner Cultural Fire Strategy' (Victorian Traditional Owner Cultural Fire Knowledge Group 2019) addresses all of this head-on by building in ongoing monitoring, evaluation and research via Principle 4 (Victorian Traditional Owner Cultural Fire Knowledge Group 2019: 7) and an explicit vision whereby science can support cultural burning 'capable of adjusting to climate change and its many challenges that are beginning to be seen and felt across the country' (Victorian Traditional Owner Cultural Fire Knowledge Group 2019: 24). There is also clear acknowledgement that factors such as the introduction of new species and the build-up of fuel across the landscape will need dedicated strategies before a new equilibrium can be reached through the reintroduction of cultural burning.

While the intergenerational knowledge, and the fundamental importance of, Country to GKLaWAC, other RAPs and Traditional Owners is now making its way into Victorian policy and procedure, more work remains to be done. The current version of the Victorian Emergency Operations Handbook (Emergency Management Victoria 2021)—the core document and reference point of emergency management in Victoria—makes explicit the importance of Aboriginal cultural heritage. It further stresses the need to minimise impacts and to report any damage caused or observed to Aboriginal cultural heritage during an emergency or its aftermath (Emergency Management Victoria 2021: 66–67). However, it stops short of requiring the presence of relevant Aboriginal RAP representatives, or indeed mandating the presence of anyone with specialist cultural knowledge at all. This not only continues the longstanding institutional disconnect between governmental policies of land management and Aboriginal connection to Country, but would also seem to require only a variable and potentially *ad hoc* recording of cultural heritage. This limited approach to documentation continues to promulgate the very obstacles currently faced when dealing with understanding the true distribution of sites across the landscape and caring for Country, as discussed at length in this chapter. We stress again that the sites registered in the VAHR are a tiny fraction of the true number of GunaiKurnai cultural heritage sites across the landscape, and that there will usually be many unregistered sites in an area burnt or threatened by wildfires.

In a discussion on the importance of addressing increasing wildfire risk to cultural heritage, Hambrecht and Rockman (2017: 637) stated that:

> Irrevocable losses of cultural heritage and key environmental and archaeological data due to the effects of anthropogenic climate change are already taking place. Future generations will judge us harshly if we do not engage seriously and effectively to save our 'burning libraries of the past' (McGovern 2016).

As impassioned as this is, the archaeological and cultural sites of GunaiKurnai country are not reducible to datasets or libraries. They are the story places of the lived experiences of more than a thousand generations of GunaiKurnai ancestors, whose Country fundamentally enfolds contemporary GunaiKurnai and will do so into the future. It was, is, and will be the responsibility of GunaiKurnai to care for Country in the face of wildfires, including when fighting wildfires with cultural burns, but always Right Fire, Right Time, Right Way and for the right reasons according to lore.

References

Aboriginal Affairs Victoria 2003. Aboriginal heritage protection and management after the 2002–2003 bushfire season: An interim report. Unpublished report, Aboriginal Affairs Victoria, Melbourne.

Abrams, M.D. and G.J. Nowacki 2008. Native Americans as active and passive promoters of mast and fruit trees in the eastern USA. *The Holocene* 18: 1123–1137.

Adeleye, M.A., S.G. Haberle, S.E. O'Connor, J. Stevenson and D.M.J.S. Bowman 2021. Indigenous fire-managed landscapes in Southeast Australia during the Holocene—New insights from the Furneaux Group islands, Bass Strait. *Fire* 4(17). DOI: 10.3390/fire4020017.

Agriculture Victoria 2022. East Gippsland National and State Parks Public Lands Map http://vro.agriculture.vic.gov.au/dpi/vro/vrosite.nsf/pages/map_eg_parks Accessed 9 February 2022.

AIDR [Australian Institute for Disaster Resilience] 2020. Major Incidents Report 2019–2020. https://knowledge.aidr.org.au/media/8049/aidr_major-incidents-report_2019-20.pdf.

Akerman, K. 1979. Heat and lithic technology in the Kimberleys, W.A. *Archaeology and Physical Anthropology in Oceania* 14(2): 144–151.

Aldeias, V., H.L. Dibble, D. Sandgathe, P. Goldberg and S.J.P. McPherron 2016. How heat alters underlying deposits and implications for archaeological fire features: A controlled experiment. *Journal of Archaeological Science* 67: 64–79.

Aldeias, V., S. Gur-Arieh, R. Maria, P. Monteiro and P. Cura 2019. *Shell* we cook it? An experimental approach to the microarchaeological record of shellfish roasting. *Archaeological and Anthropological Sciences* 11: 389–407.

Alexander, M.E. and M.G. Cruz 2011. Interdependencies between flame length and fireline intensity in predicting crown fire initiation and crown scorch height. *International Journal of Wildland Fire* 21(2): 95–113.

Alexandra, J. and C.M. Finlayson 2020. Floods after bushfires: rapid responses for reducing impacts of sediment, ash, and nutrient slugs. *Australasian Journal of Water Resources* 24(1): 9–11.

Ancel, B. and V. Py 2008. L'abattage par le feu: Une technique minière ancestrale. *Archéopages: Mines et Carrières* 22: 34–41.

Andersson, R. 2005. Historical Land-use Information from Culturally Modified Trees. Unpublished PhD dissertation, Swedish University of Agricultural Sciences.

Andrus, C.F.T. and D.E. Crowe 2002. Alteration of Otolith Aragonite: Effects of Prehistoric Cooking Methods on Otolith Chemistry. *Journal of Archaeological Science* 29: 291–299.

Anonymous 1869. Ella's dream of how the trees of Nuntin Forest died. *Gippsland Times* 3 August 1869: 4.

Anonymous 1891. Gippsland forests. *Great Southern Advocate*, 22 May 1891: 4.

Anonymous 1905. Reminiscences of early Gippsland. *Gippsland Times*, 2 February 1905: 3.

Anonymous. 1931. A baobab as lock-up. *The Queenslander* 26 February: 29.

Asmussen, B. 2009. Intentional or incidental thermal modification? Analysing site occupation via burned bone. *Journal of Archaeological Science* 36: 528–536.

Baby, R.S. 1954. *Hopewell Cremation Practices*. Papers in Archaeology, number 1, Ohio Historical Society.

Backhouse, J. 1843. *A Narrative of a Visit to the Australian Colonies* (Vol. 11). London: Hamilton, Adams.

Bahr, H.A., G. Fischer and H.J. Weiss 1986. Thermal-shock crack patterns explained by single and multiple crack propagation. *Journal of Materials Science* 21(8): 2716–2720.

Baker, A.G., C. Catterall and K. Benkendorff 2021. Invading rain forest pioneers initiate positive fire suppression feedbacks that reinforce shifts from open to closed forest in eastern Australia. *Journal of Vegetation Science* 32(6): e13102. DOI: 10.1111/jvs.13102.

Banks, J. 1997. Trees: the silent fire historians. *Bogong* 18: 9–12.

Bennett, J.L. 1999. Thermal alteration of buried bone. *Journal of Archaeological Science* 26: 1–8.

Binford, L.R. 1963. An analysis of cremations from three Michigan sites. *Wisconsin Archaeologist* 44(2): 98–110.

Binford, L.R. and J.F. O'Connell 1984. An Alyawarra Day: The Stone Quarry. *Journal of Anthropological Research* 40(3): 406–432.

Bird, E.C.F. 1966. The impact of man on the Gippsland lakes, in S.R. Eyre and G.R. Jones (eds) *Geography as human ecology: Methodology by example*: 55–73. London: Edward Arnold.

Black, L. 1941. *Burial Trees*. Melbourne: Robertson and Mullens Ltd.

Blundell, V. and D. Woolagoodja 2005. *Keeping the Wanjinas fresh*. Fremantle: Fremantle Arts Centre Press.

Bohemia, J. and W. McGregor 1995. *Nyibayarri: Kimberley Tracker*. Canberra: Aboriginal Studies Press.

Bonneau, A., F. Brock, T. Higham, D.G. Pearce and A.M. Pollard 2011. An improved pretreatment protocol for radiocarbon dating black pigments in San rock art. *Radiocarbon* 53(3): 419–428.

Bowler, J.M., H. Johnston, J.M. Olley, J.R. Prescott, R.G. Roberts, W. Shawcross and N.A. Spooner 2003. New ages for human occupation and climatic change at Lake Mungo, Australia. *Nature* 421(6925): 837–840.

Bowman, D. M., B.P. Murphy, D.L. Neyland, G.J. Williamson and L.D. Prior 2014. Abrupt fire regime change may cause landscape-wide loss of mature obligate seeder forests. *Global Change Biology* 20(3): 1008–1015.

Bradstock, R.A. and T.D. Auld 1995. Soil temperatures during experimental bushfires in relation to fire intensity: consequences for legume germination and fire management in south-eastern Australia. *Journal of Applied Ecology* 76–84.

Brain, C.K. 1983. *The Hunters or the Hunted? An Introduction to African Cave Taphonomy*. Chicago: University of Chicago Press.

Brian, R. 1979. *The decorated body*. London: Hutchison.

Bride, T.F. (ed.) 1898. *Letters from Victorian pioneers: Being a series of papers on the early occupation of the colony, the aborigines, etc.* Melbourne: Robert S. Brain.

Brodard, A., P. Guibert, C. Ferrier, É. Debard, B. Kervazo and J.-M. Geneste 2014. Les rubéfactions des parois de la grotte Chauvet: Une histoire de chauffe? in P. Paillet (ed.) *Les Arts de la Prehistoire: Micro-analyses, Mises en Contextes et Conservation*: 233–235. Les-Eyzies-de-Tayac: Musée National de Préhistoire.

Brodribb, W. 1976. 1883. *Recollections of an Australian squatter*. Melbourne: Queensberry Hill Press.

Brooks, A.S., P.E. Hare, J.E. Kokis and K. Durana 1991. A burning question: Differences in laboratory induced and natural diagenesis in ostrich eggshell proteins. *Annual Report of the Geophysical Laboratory* 2250: 176–179. Carnegie Institute of Washington.

Buckley, P.C. 1844–1872. Patrick Coady Buckley Journal, Royal Historical Society of Victoria, MS 000097, Box 37/4.

Buenger, B.A. 2003. *The impact of wildland and prescribed fire on archaeological resources.* Unpublished PhD Dissertation, University of Kansas.

Buikstra, J.E. and M. Swegle 1989. Bone modification due to burning: Experimental evidence, in R. Bonnichsen and H. Sorg (eds) *Bone Modification*: 247–258. Orono: Center for the Study of the First Americans, Institute of Quaternary Studies, University of Maine.

Burchett, C. No date. 'A Diary and Letters of Caleb Burchett covering his life from birth in 1843 to about 1900; together with notes and letters from his father James, and his brother James', MS 8814, State Library of Victoria.

Bushfire Recovery Victoria 2020. Eastern Victorian Fires 2019–2020 State Recovery Plan. Melbourne: Bushfire Recovery Victoria.

Butler, B.W. and J.D. Cohen 1998. Firefighter safety zones: A theoretical model based on radiative heating. *International Journal of Wildland Fire* 8(2): 73–77.

Cahir, F. and S. McMaster 2018. Fire in Aboriginal south-eastern Australia, in I.D. Clark and P.A. Clarke (eds), *Aboriginal biocultural knowledge in south-eastern Australia: Perspectives of early colonists*: 115–132. Melbourne: CSIRO Publishing.

Callender, W.R., E.N. Powell and G.M. Staff 1994. Taphonomic Rates of Molluscan Shells Placed in Autochthonous Assemblages on the Louisiana Continental Slope. *Palaios* 9: 60–73.

Canadell, J.G., C.P. Meyer, G.D. Cook, A. Dowdy, P.R. Briggs, J. Knauer, A. Pepler and V. Haverd 2021. Multi-decadal increase of forest burned area in Australia is linked to climate change. *Nature Communications* doi:10.1038/s41467-021-27225-4.

Canning, S. 2003. *Archaeology, Cultural Resource Management, Planning and Predictive Modelling in the Melbourne Metropolitan*. Unpublished PhD dissertation, La Trobe University.

Castillo, R.F., D.H. Ubelaker, J.A.L. Acosta, and G.A.C. de la Fuente 2013. Effects of temperature on bone tissue: Histological study of the changes in the bone matrix. *Forensic Science International* 226: 33–37.

CFA [Country Fire Authority]. 2011. *Fire Ecology: Guide to Environmentally Sustainable Bushfire Management in Rural Victoria.* https://www.cfa.vic.gov.au/ArticleDocuments/550/FireEcologyGuide_Final_web.pdf.aspx?Embed=Y.

Chakrabarti, B., T. Yates and A. Lewry 1995. Effects of fire damage on natural stonework in buildings. *Construction and Building Materials* 10(7): 539–544.

Chalmin, E. and J. Huntley 2018. Rock Art Pigment Characterisation, in B. David and I. McNiven (eds), *Oxford Handbook of the Archaeology and Anthropology of Rock Art.* Oxford: Oxford University Press. DOI: 10.1093/oxfordhb/9780190607357.001.0001.

Chalmin, E., C. Vignaud, F. Farges and M. Menu 2008. Heating effect on manganese oxihydroxides used as black Palaeolithic pigment. *Phase Transitions* 81(2-3): 179–203.

Chambers, D.P. and P.M. Attiwill 1994. The Ash-Bed Effect in *Eucalyptus regnans* Forest: Chemical, Physical and Microbiological Changes in Soil After Heating or Partial Sterilisation. *Australian Journal of Botany* **42**: 739–749.

Clark, I.D. 1994. George Augustus Robinson's 1844 Journey through Gippsland. *Gippsland Heritage Journal* 17: 12–19.

Clarke, J. and N. North 1991. Chemistry of deterioration of post-estuarine rock art in Kakadu National Park, in C. Pearson and B.K. Swarts Jr (eds) *Rock art and posterity: conserving, managing and recording rock art*, 4: 88–92. Melbourne: Occasional AURA Publications, Australian Rock Art Association.

Cole, N.A. and L.A. Wallis 2019. Indigenous rock art tourism in Australia: Contexts, trajectories, and multifaceted realities. *Arts* 8(4): 162.

Cook, N., I. Davidson and S. Sutton 1990. Why are so many ancient rock paintings red? *Australian Aboriginal Studies* (1): 30–32.

Coombe, R.J. and M.F. Baird 2015. The limits of heritage: corporate interests and cultural rights on resource frontiers, in W. Logan, M.N. Craith and U. Kockel (eds) *A Companion to Heritage Studies: 337-354*. Hoboken: John Wiley and Sons.

Correia, P.M. 1997. Fire modification of bone: A review of the literature, in W.D. Haglund and M.H. Sorg (eds) *Forensic Taphonomy: The Postmortem Fate of Human Remains*: 275–293. Boca Raton: CRC Press.

Coutts, P.J.F., R.K. Frank and P. Hughes 1978. Aboriginal Engineers of the Western District, Victoria. *Records of The Victorian Archaeological Survey* 7. Ministry for Conservation, Melbourne.

Crisp, M. 2013. Amino acid racemisation dating: Method development using African ostrich (*Struthio camelus*) eggshell. Unpublished PhD dissertation, University of York.

Currey, J.D. 1979. The effect of drying on the strength of mollusc shells. *Journal of the Zoological Society of London* 188: 301–388.

Currey, J.D. 1988. Shell Form and Strength, in E.R. Trueman and M.R. Clarke (eds) *Form and Function, The Mollusca* 11: 183-210. San Diego and London: Academic Press.

David, B. 1990. How was this bone burnt? in S. Solomon, I. Davidson and D. Watson (eds) Problem Solving in Taphonomy: Archaeological and Palaeontological Studies from Europe, Africa and Oceania. *Tempus* 2: 65–79.

David, B., J.-J. Delannoy, R. Gunn, L.M. Brady, F. Petchey, J. Mialanes, E. Chalmin, J.-M. Geneste, I. Moffat, K. Aplin and M. Katherine 2017. Determining the age of paintings at JSARN–113/23, Jawoyn Country, central-western Arnhem Land plateau, in B. David, P. Taçon, J.-J. Delannoy and J.-M. Geneste (eds) *The Archaeology of Rock Art in Western Arnhem Land, Australia*: 371–422. Canberra: ANU Press.

David, B., J. Fresløv, R. Mullett, GunaiKurnai Land and Waters Aboriginal Corporation, J.-J. Delannoy, M. McDowell, C. Urwin, J. Mialanes, F. Petchey, R. Wood, L. Russell, L.J. Arnold, B. Stephenson, R. Fullagar, J. Crouch, J. Ash, J. Berthet, V.N.L. Wong and H. Green 2021. 50 years and worlds apart: Rethinking the Holocene occupation of Cloggs Cave (East Gippsland, SE Australia) five decades after its initial archaeological excavation and in light of GunaiKurnai world views. *Australian Archaeology* 87: 1–20.

David, B., L. Lamb, J.-J. Delannoy, F. Pivoru, C. Rowe, M. Pivoru, T. Frank, N. Frank, A. Fairbairn and R. Pivoru 2012. Poromoi Tamu and the case of the drowning village: History, lost places and the stories we tell. *International Journal of Historical Archaeology* 16: 319–345.

David, B., J. Mialanes, F. Petchey, K. Aplin, J.-M. Geneste, R. Skelly and C. Rowe 2015. Archaeological investigations at Waredaru and the origins of the Keipte Kuyumen clan estate, upper Kikori River, Papua New Guinea. *Paleo* 26: 33–57.

David, M., R. Mullett and R. Mullett 1998. The Caledonia fire area, Alpine National Park, Victoria: An archaeological survey and Aboriginal heritage assessment. Unpublished report to Aboriginal Affairs Victoria, Melbourne.

Davies, R. 1846. On the Aborigines of Van Diemen's Land. *Tasmanian Journal of Natural Science* 2: 409–420.

Dayet, L., F.X. Le Bourdonnec, F. Daniel, G. Porraz and P.J. Texier 2016. Ochre provenance and procurement strategies during the middle stone age at Diepkloof Rock Shelter, South Africa. *Archaeometry* 58(5): 807–829.

Deal, K. 2012. Fire Effects on Flaked Stone, Ground Stone, and Other Stone Artifacts, in K. Ryan, A. Jones, C. Koerner and K. Lee (eds.) *Wildland Fire in Eco-systems: Effects of Fire on Cultural Resources and Archaeology*: 97–111. General Technical Report RMRS-GTR-42- Vol. 3. Fort Collins (CO): U.S. Department of Agriculture, Rocky Mountain Research Station.

Delannoy J.-J., B, David, J. Fresløv, R. Mullett, R., GunaiKurnai Land and Waters Aboriginal Corporation, H. Green, J. Berthet, F. Petchey, L.J. Arnold, R. Wood, M. McDowell, J. Crouch, J. Mialanes, J. Ash and V.N.L. Wong 2020. Geomorphological context and formation history of Cloggs Cave: What was the cave like when people inhabited it? *Journal of Archaeological Science: Reports* 33: 102461. DOI: 10.1016/j.jasrep.2020.102461.

Demarchi, B., M.G. Williams, N. Milner, N. Russell, G. Bailey and K.E.H. Penkman 2011. Amino acid racemization dating of marine shells: a mound of possibilities. *Quaternary International* 239: 114–124.

Department of Primary Industries and Parks Victoria 2017. Timbarra fire complex Gippsland region, Bushfire Rapid Assessment Team report. Unpublished report to the State Government of Victoria, Melbourne.

Devlin, J.B. and N.P. Herrmann 2008. Bone color as an interpretive tool of the depositional history of archaeological cremains, in C.W. Schmidt and S.A. Symes (eds) *The Analysis of Burned Human Remains*: 109–x. Academic Press. DOI: 10.1016/B978-0-12-372510-3.X5001-1.

Dibden J. 2019. *Drawing in the Land: Rock Art in the Upper Nepean, Sydney Basin, New South Wales.* Canberra: ANU Press.

Douglas, R. 1981. *Exploration and Settlement of the Bairnsdale District From 1770 to 1870—The First Hundred Years.* Bairnsdale: Bairnsdale Teachers' Centre/Mitchell River Press.

Dow, C. 2004. Tatungalung Country: An environmental history of the Gippsland lakes. Unpublished PhD dissertation, Monash University.

Driscoll, K. and J. Menuge 2011. Recognising burnt vein quartz artefacts in archaeological assemblages. *Journal of Archaeological Science* 38: 2251–2260.

Eggerath, H. 2003. Die Geschichte des Neandertals, in B. Baumgärtel and K. Thelen (eds.) *Bewegte Landschaften: Der Düsseldorfer Malerschule*: 60–73. Heidelberg: Edition Braus.

Elkin, A.P. 1948. Pressure Flaking in the Northern Kimberley, Australia. *Man* 130: 110–113.

Elms, A. W. 1920. A Fiery Summer, in *The Land of the lyre bird: A story of early settlement in the great forest of South Gippsland*: 305–313. Melbourne: Gordon and Gotch.

Emergency Management Victoria 2021. Victorian Emergency Operations Handbook. Edition 4—November 2021. Melbourne: Emergency Management Victoria.

Emory, K.P. 1975. *Material Culture of the Tuamotu Archipelago.* Honolulu: Bishop Museum.

Ericsson, T.S., L. Östlund and R. Andersson, R. 2003. Destroying a path to the past—the loss of culturally scarred trees and change in forest structure along Allmunvägen, in mid-west boreal Sweden. *Silva Fennica* 37(2): 283–298.

Etheridge, R. 1918. *The Dendroglyphs or 'Carved Trees' of New South Wales.* Sydney: William Applegate Gullick, Government Printer.

Faust, G.T. 1950. Thermal analysis studies on carbonates: Aragonite and calcite. *American Mineralogist* 35: 207–244.

Fell, L. 1978. Changes in the Gippsland Lakes environment over seventy years. *Clematis* 17: 13–15.

Fels, M. 2011. *I Succeeded Once: The Aboriginal Protectorate on the Mornington Peninsula, 1839-1840.* Canberra: ANU Press.

Fernandes, P.M. and H.S. Botelho 2003. A review of prescribed burning effectiveness in fire hazard reduction. *International Journal of Wildland Fire* 12: 117–128.

Fernández-Jalvo, Y., L. Tormo, P. Andrews and M.D. Marin-Monfort 2018. Taphonomy of burnt bones from Wonderwerk cave (South Africa). *Quaternary International* 495: 19–29.

Fiers, G., É. Halbrucker, T.D. Kock, H. Vandendriessche, P. Crombé and V. Cnudde 2021. Thermal Alteration of Flint: An Experimental Approach to Investigate the Effect on Material Properties. *Lithic Technology* 46: 1–18.

Fison, L. and A.W. Howitt 1880. *Kamileroi and Kurnai: Group marriage and relationship, and marriage by elopement drawn chiefly from the usage of the Australian aborigines; also the Kurnai tribe; their customs in peace and war*. Melbourne: George Robertson.

Florek, S. 1989. Fire in the Quarry. *Australian Archaeology* 29: 22–27.

Ford, J. 2006. Painting Contact: Characterising the Paints of the South Woronora Plateau Rock Art Assemblage, Wollongong, New South Wales. Unpublished BA(Hons) dissertation, The Australian National University.

Freeman, D., B. Williamson and J. Weir 2021. Cultural burning and public sector practice in the Australian Capital Territory. *Australian Geographer* 52(2): 111–129.

Fresløv, J. 2004. Post Wildfire Indigenous Heritage Survey, Volume 2: Management of Impacts from Wildfire and Suppression Activities. Unpublished report to Parks Victoria, the Department of Sustainability, and Aboriginal Affairs Victoria. Perspectives Heritage Solutions Pty Ltd, Hurstbridge.

Fresløv, J. 2022. The Lakes National Park, Gippsland Lakes Coastal Park and Raymond Island Reserve, Cultural Heritage Survey Stage 3. Unpublished report to Parks Victoria and the GunaiKurnai Traditional Owner Land Management Board, Kalimna West.

Fresløv J., P. Hughes and R. Mullett 2004. Post Wildfire Indigenous Heritage Survey. Volume 1: Background, Survey, Results and Recommended Management Options. Unpublished cultural heritage management report to Aboriginal Affairs Victoria, Parks Victoria and the Department of Sustainability and Environment, Melbourne, Victoria.

Fresløv, J. and Mullett, R. in press. Below the sky, above the clouds: The archaeology of the Australian High Country. In B. David and I. J. McNiven (eds), *The Oxford Handbook of the Archaeology of Indigenous Australia and New Guinea*. Oxford: Oxford University Press.

Gaffey, S.J., J.J. Kolak and C.E. Bronnimann 1991. Effects of drying, heating, annealing, and roasting on carbonate skeletal material, with geochemical and diagenetic implications. *Geochimica et Cosmochimica Acta* 55: 1627–1640.

Gammage, B. 2011. *The Biggest Estate on Earth: How Aborigines made Australia*. Sydney: Allen and Unwin.

Gaughwin, D. 1981. Sites of Archaeological Significance in the Western Port Catchment; a Draft, October 1981. Unpublished report prepared for Ministry for Conservation.

Gell, P.A., I.-M. Stuart and J.D. Smith 1993. The response of vegetation to changing fire regimes and human activity in East Gippsland, Victoria, Australia. *The Holocene* 3(2): 150–160.

Germinario, L., G.F. Andriani and R. Laviano 2015. Decay of calcareous building stone under the combined action of thermoclastism and cryoclastism: A laboratory simulation. *Construction and Building Materials* 75: 385-39.

Gibson, J.M. and R. Mullett 2020. The last Jeraeil of Gippsland: Rediscovering an Aboriginal ceremonial site. *Ethnohistory* 67(4): 551–577.

Gibson, J.M. and R. Mullett 2021. After 140 years, researchers have rediscovered an important Aboriginal ceremonial ground in East Gippsland. *The Conservation* 5 May.

Gilchrist, R. and H.C. Mytum 1986. Experimental archaeology and burnt animal bone from archaeological sites. *Circaea* 4: 29–38.

Glover, C.P. and S.M. Kidwell 1993. Influence of Organic Matrix on the Post-Mortem Destruction of Molluscan Shells. *Journal of Geology* 101: 727–747.

Godfrey-Smith, D.I. and S. Ilani 2004. Past thermal history of goethite and hematite fragments from Qafzeh Cave deduced from thermal activation characteristics of the 110 °C TL peak of enclosed quartz grains. *ArchéoSciences, revue d'Archéométrie* 28(1): 185–190.

González-Gómez, W.S., P. Quintana, A. May-Pat, F. Avilés, J. May-Crespo and J.J. Alvarado-Gil 2015. Thermal effects on the physical properties of limestones from the Yucatan Peninsula. *International Journal of Rock Mechanics and Mining Science* 75: 182–189.

Gould, R.A. 1976. A Case of Heat Treatment of Lithic Materials in Aboriginal Northwestern California. *The Journal of California Anthropology* 3: 142–144.

Graham, D. 2020. In Fox Koob, S. I couldn't let it burn: Buchan couple saved precious picture before home turned to ash, *The Age*, 13 January. Melbourne.

Gregg, M.L. and R.J. Grybush 1976. Thermally Altered Siliceous Stone from Prehistoric Contexts: Intentional versus Unintentional Alteration. *American Antiquity* 41: 189–192.

Griffiths, B. and L. Russell 2018. What we were told: Responses to 65,000 years of Aboriginal history. *Aboriginal History Journal* 42: 31–54.

Griffiths, T. 2002. Judge Stretton's fires of conscience. *Gippsland Heritage Journal* 26: 9-18.

Guérard, E. von. 1870. Reply on the critic of Eugène von Guérard's painting of the North Grampians, James Smith Papers, Mitchell Library, State Library of New South Wales. MS 212/4.

Guler, S., Z.F. Türkmenoğlu and O.O. Varol 2021. Thermal shock and freeze-thaw resistance of different types of carbonate rocks. *International Journal of Rock Mechanics and Mining Sciences* 137: 104545. DOI: 10.1016/j.ijrmms.2020.104545.

Gunaikurnai Land and Waters Aboriginal Corporation. 2015. Gunaikurnai Whole-of-Country Plan. https://gunaikurnai.org/wp-content/uploads/2021/07/Gunaikurnai-Whole-of-Country-Plan-ONLINE.pdf.

Haberle, S. G., S. Rule, P. Roberts, H. Heijnis, G. Jacobsen, C. Turney, R. Cosgrove, A. Ferrier, P. Moss, S. Mooney and P. Kershaw **2010.** Paleofire in the wet tropics of northeast Queensland, Australia. *PAGES News* 18(2): 78–79.

Haecker, C. 2012. Chapter 6: Fire Effects on Material of the Historic Period, in K.C. Ryan, A. Trinkle, C.L. Koerner, K.M. Lee (eds) *Wildland Fire in Ecosystems: Effects of Fire on Cultural Resources and Archaeology*: 131–142. General Technical Report RMRS-GTR 42, volume 3. Fort Collins (CO): US Department of Agriculture, Forest Service, Rocky Mountain Research Station.

Hall, K., I. Meiklejohn and J. Arocena 2007. The thermal responses of rock art pigments: implications for rock art weathering in southern Africa. *Geomorphology* 91(1): 132–145.

Hall, K. and C.E. Thorn 2014. Thermal fatigue and thermal shock in bedrock: An attempt to unravel the geomorphic processes and products. *Geomorphology* 206: 1–13.

Hambrecht, G. and M. Rockman 2017. International Approaches to Climate Change and Cultural Heritage. *American Antiquity* 82: 627–641.

Hancock, W.K. 1972. *Discovering Monaro: A Study of Man's Impact on his Environment*. Cambridge: Cambridge University Press.

Hanein, T., M. Simoni, C.L. Woo, J.L. Provis and H. Kinoshita 2021. Decarbonisation of calcium carbonate at atmospheric temperatures and pressures, with simultaneous CO_2 capture, through production of sodium carbonate. *Energy & Environmental Science* 14: 6595–6604.

Harrisson, T. 1957. Niah Workbook—March 1957 book 5. Unpublished field notebook. Harrisson Excavation Archive. Kuching, Sarawak Museum.

Head, L., D. Trigger and J. Mulcock 2005. Culture as concept and influence in environmental research and management. *Conservation and Society* 3(2): 251–264.

Henry, D.O., C. Cordova, J.J. White, R.M. Dean, J.E. Beaver, H. Ekstrom, S. Kadowaki, J. McCorriston, A. Nowell and L. Scott-Cummings 2003. The Early Neolithic Site of Ayn Abū Nukhayla, Southern Jordan. *Bulletin of the American School of Oriental Research* 330: 1–30.

Herrmann, N.P. and J.L. Bennett 1999. The differentiation of traumatic and heat-related fractures in burned bone. *Journal of Forensic Science* 44: 461–469.

Hiscock, P. 2007. Looking the other way. A materialist/technological approach to classifying tools and implements, cores and retouched flakes, in S. McPherron (ed.) *Tools versus cores? alternative approaches to stone tool analysis*: 198–222. Newcastle: Cambridge Scholars Publishing.

Hoerlé, S. 2006. Rock temperatures as an indicator of weathering processes affecting rock art. *Earth Surface Processes and Landforms* 31(3): 383–389.

Hollis, J.J., J.S. Gould, M.G. Cruz and W. Lachlan McCaw 2015. Framework for an Australian fuel classification to support bushfire management. *Australian Forestry* 78: 1–17.

Holtz, D., A. Markham, K. Cell and B. Ekwurzel 2014. *National Landmarks at Risk: How Rising Seas, Floods, and Wildfires Are Threatening the United States' Most Cherished Historic Sites*. Union of Concerned Scientists.

Howitt, A.W. 1860. Letter to his mother. Howitt Papers: State Library of Victoria, 1045/ 3a, nos. 10 and 11.

Howitt, A.W. 1890. The eucalypts of Gippsland. *Transactions of the Royal Society of Victoria* 2: 81–120.

Howitt, A.W. 1904. *The Native Tribes of South-East Australia*. London: Macmillan & Co.

Howitt, A.W. and L. Fison, No date. John Campbell MacLeod archive. https://howittandfison.org/article/61750. Accessed: 18 September 2021.

Hughes, R. and D. Mercer 2009. Planning to reduce risk: the wildfire management overlay in Victoria, Australia. *Geographical Research* 47: 124–141.

Humboldt, A. von. 1845–1862. *Kosmos: Entwurf einer physischen Weltgeschichte*. Stuttgart and Tübingen: J.G. Cotta'schen Buchhandlung (1–5).

Humboldt, A. von. 1849. *Views of Nature*. Trans. Mrs Sabine. London: Longman, Brown, Green and Longmans, and J. Murray.

Humboldt, A. von. 1849–1858. *Cosmos: A Sketch of a Physical Description of the Universe*. Trans. E.C. Otté. London: Henry G. Bohn (1–3).

Huntley. J. 2019. Appletree Aboriginal Area Plan of Management. Unpublished report to the NSW National Parks and Wildlife Service (Bulga Office), Office of Environment and Heritage and the Wonnarua Advisory Committee.

Huntley, J. 2021 Australian Indigenous Ochres: Use, Sourcing and Exchange. In I. McNiven and B. David (Eds), *The Oxford Handbook of the Archaeology of Indigenous Australia and New Guinea*. Oxford: Oxford University Press. DOI: 10.1093/oxfordhb/9780190095611.013.21.

Huntley, J., M. Aubert, A.A. Oktaviana, R. Lebe, B. Hakim, B. Burhan, L.M. Aksa, I.M. Geria, M. Ramli, L. Siagian, H.E.A. Brand and A. Brumm 2021. The effects of climate change on the Pleistocene rock art of Sulawesi. *Scientific Reports* 11 (9833) DOI: 10.1038/s41598-021-87923-3.

Huntley, J., M. Aubert, J. Ross, H.E.A. Brand and M.J. Morwood 2015. One colour, (at least) two minerals: a study of mulberry rock art pigment and a mulberry pigment 'quarry' from the Kimberley, northern Australia. *Archaeometry* 57(1): 77–99.

Huntley, J., H. Brand, M. Aubert and M. Morwood 2014. The first Australian Synchrotron powder diffraction analysis of pigment from a Wandjina motif in the Kimberley, Western Australia. *Australian Archaeology* 78(1): 33–38.

Huntley, J., S. George, M.J. Sutton and P. Taçon 2018. Second-hand? Insights into the age and 'authenticity' of colonial period rock art on the Sunshine Coast, Queensland, Australia. *Journal of Archaeological Science: Reports* 17: 163–172.

Huntley J. and K. Officer 2016. pXRF and SEM analysis of dusts on the rock art panels for Castle Rock. Unpublished report, Navin Officer Heritage Consulting, for Wilpinjong Coal Pty Ltd.

Hynes, R.A. and A.K. Chase 1982. Plants, sites and domiculture: Aboriginal influence upon plant communities in Cape York Peninsula. *Archaeology in Oceania* 17: 38–50.

IPCC 2022. Climate Change 2022: Impacts, Adaptation and Vulnerability. Contribution of Working Group II to the Sixth Assessment Report of the Intergovernmental Panel on Climate Change [H.-O. Pörtner, D.C. Roberts, M. Tignor, E.S. Poloczanska, K. Mintenbeck, A. Alegría, M. Craig, S. Langsdorf, S. Löschke, V. Möller, A. Okem, B. Rama (eds.)]. Cambridge University Press. Cambridge University Press, Cambridge, UK and New York, NY, USA, 3056 pp., doi:10.1017/9781009325844.

Johnson, L.B., E. Larson, K.G. Gill and J. Savage 2022. Blending tree-ring fire-scar records and indigenous memory in northern Minnesota, USA. *Pages Magazine, Special Section: Socio-Ecological Approaches to Conservation* 30(1): 36–37.

Jones, R. 1969. Firestick farming. *Australian Natural History* 16: 224–228.

Joubert, J. 1876. The Monaro district. *Australian Town and Country Journal*, 2 November, p. 20.

Jurskis, V. 2015. *Firestick Ecology: Fairdinkum Science in Plain English*. Ballarat: Connor Court Publishing.

Keeley, J.E. 2009. Fire intensity, fire severity and burn severity: a brief review and suggested usage. *International journal of Wildland Fire* 18: 116–126.

Kelly, T. 2003. Northeast fire recovery assessment: Aboriginal cultural heritage impact assessment, scoping study–draft report (cultural heritage investigation: field report). Unpublished report to Parks Victoria, Melbourne.

Khan, W.N. and R. Chhibber 2021. Utilization of red ochre in developing welding electrodes for offshore welds. *Materials Today: Proceedings* 41: 870–873.

Kharuk, V.I., M.L. Dvinskaya, S.T. Im, A.S. Golyukov and K.T. Smith 2022. Wildfires in the Siberian Arctic. *Fire* doi:10.3390/fire5040106.

Knight, J.A. 1985. Differential preservation of calcined bones at the Hirundo site, Alton, Maine. Unpublished M.Sc. dissertation, University of Maine.

Koon, H.E.C., R.A. Nicholson and M.J. Collins 2003. A practical approach to the identification of low temperature heated bone using TEM. *Journal of Archaeological Science* 30: 1393–1399.

Koppel, B., K. Szabó, M.W. Moore and M.J. Morwood 2017. Untangling time-averaging in shell middens: Defining temporal units using amino acid racemisation. *Journal of Archaeological Science: Reports* 7: 741–750.

Kozuch, L. 2003. Use of fire in shell bead manufacture at Cahokia. *Bulletin of the Florida Museum of Natural History* 44: 81–90.

Lacanette, D., J.C. Mindeguia, A. Brodard, C. Ferrier, P. Guibert, J.C. Leblanc, P. Malaurent and C. Sirieix 2017. Simulation of an experimental fire in an underground limestone quarry for the study of Paleolithic fires. *International Journal of Thermal Sciences* 120: 1–18.

Lambert, D. and B. Welsh 2011. Fire and rock art. *Rock Art Research* 28(1): 45–48.

Larsen, S.C. 2015. Recrystallization of biogenic aragonite shells from archaeological contexts and implications for paleoenvironmental reconstruction. Unpublished M.A dissertation, Western Washington University.

Lees, E.H. 1914a. Forests and fires. *The Australasian*, 24 October 1914: 8.

Lees, E.H. 1914b. Letter to the editor. *The Australasian*, 14 November 1914: 13.

Lees, E.H. 1915. What is Nardoo? *Victorian Naturalist* 31(9): 133–135.

Legodi, M.A. and D. de Waal 2007. The preparation of magnetite, goethite, hematite and maghemite of pigment quality from mill scale iron waste. *Dyes and Pigments* 74(1): 161–168.

Lehman, G. 2000. Turning back the clock: Fire, biodiversity, and Indigenous community development in Tasmania, in R. Baker, J. Davies and E. Young (eds.) *Working on Country: Contemporary Indigenous Management of Australia's Lands and Coastal Regions*: 308–319. Oxford: Oxford University Press.

Lindauer, S., S. Milano, A. Steinhof and M. Hinderer 2018. Heating mollusc shell—A radiocarbon and microstructure perspective from archaeological shells recovered from Kalba, Sharjah Emirate, UAE. *Journal of Archaeological Science: Reports* 21: 528–537.

Lindenmayer, D.B., R.M. Kooyman, C. Taylor, M. Ward and J.E. Watson 2020. Recent Australian wildfires made worse by logging and associated forest management. *Nature Ecology & Evolution* 4(7): 898–900.

Long, A. 2003. Scarred Trees: An Identification and Recording Manual. Unpublished manual. Melbourne: Aboriginal Affairs.

MacDonald, B.L., R.G.V. Hancock, A. Cannon, F. McNeill, R. Reimer and A. Pidruczny 2013. Elemental analysis of ochre outcrops in southern British Columbia, Canada. *Archaeometry* 55(6): 1020–1033.

MacDonald, B.L., S. David, X. He, F. Rahemtulla, D. Emerson, P.A. Dube, M.R. Maschmann, C.E. Klesner and T.A. White 2019. Hunter-Gatherers Harvested and Heated Microbial Biogenic Iron Oxides to Produce Rock Art Pigment. *Scientific Reports* 9(1): 1–13.

Macdonald, D. 1887. *Gum boughs and wattle bloom*. London: Cassell.

Marcaida, I., M. Maguregui, S. Fdez-Ortiz de Vallejuelo, H. Morillas, N. Prieto-Taboada, M. Veneranda, K. Castro and J.M. Madariaga 2017. In situ X-ray fluorescence-based method to differentiate among red ochre pigments and yellow ochre pigments thermally transformed to red pigments of wall paintings from Pompeii. *Analytical and Bioanalytical Chemistry* 409(15): 3853–3860.

Marin, F. and G. Luquet. 2004. Molluscan Shell Proteins. *Comptes Rendus Palevol: General Palaeontology (Palaeobiochemistry)* 3: 469–492.

Marques, M.P.M., A.P. Mamede, A.R. Vassalo, C. Makhoul, E. Cunha, D. Gonçalves, S.F. Parker and L.A.E. Batista de Carvalho 2018. Heat-induced Bone Diagenesis Probed by Vibrational Spectroscopy. *Scentific Reports* 8: 15935. DOI: 10.1038/s41598-018-34376-w.

Marshall, M., K. May, R. Dann and L. Nulgit 2020. Indigenous Stewardship of Decolonised Rock Art Conservation Processes in Australia. *Studies in Conservation* 65(sup1): 205–212.

Matthews, S. 2013. Dead fuel moisture research: 1991–2012. *International Journal of Wildland Fire* 23: 78-92.

Mays, S. 2021. *The Archaeology of Human Bones*. Routledge, London.

McArthur, A.G. 1967. *Fire behaviour in eucalypt forests*. Canberra: Forestry and Timber Bureau.

McCutcheon, P.T. 1992. Burned archaeological bone, in J.K. Stein (ed.) *Deciphering a shell midden*: 347–370. San Diego: Academic Press.

McDowell, M.C., B. David, R. Mullett, J. Fresløv, GunaiKurnai Land and Waters Aboriginal Corporation, J.-J. Delannoy, J. Mialanes, C. Thomas, J. Ash, J. Crouch, F. Petchey, J. Buettel and L.J. Arnold. 2022. Interpreting the mammal deposits of Cloggs Cave (SE Australia),

GunaiKurnai Aboriginal Country, through community-led partnership research. *People and Nature* 4: 1629–1643.

McGovern, T.H. 2016. Endangered Environmental Archives in the North Atlantic, Crisis and Response. Paper given at the 81st Annual Meeting of the Society for American Archaeology, Orlando, Florida.

McKinley, J.I. 2000. The analysis of cremated bone, in M. Cox and S. Mays (eds) *Human Osteology in Archaeology and the Forensic Sciences*: 403–490. Cambridge: Cambridge University Press.

McLean, A. No date. Reminiscences of Allan McLean. Royal Historical Society of Victoria, MS 000384, Box 125/7.

Mercieca, A. 2000. Burnt and broken: an experimental study of heat fracturing in silcrete. *Australian Archaeology* 51: 40–47.

Mercieca, A. and P. Hiscock 2008. Experimental insights into alternative strategies of lithic heat treatment. *Journal of Archaeological Science* 35(9): 2634–2639.

Mialanes, J., David, B., Stephenson, B., Fresløv J., Mullett, R., GunaiKurnai Land and Waters Aboriginal Corporation, Metz, L., Delannoy, J.-J., Crouch, J., McDowell, M., Petchey, F., Green, H., Arnold, L. J., Wood, R., Ash, J. in press. The *mulla-mullung*'s *bulk*: GunaiKurnai perspectives on the stone artefacts of Cloggs Cave, GunaiKurnai Country, southeastern Australia. In C. Smith (ed.), *The Oxford Handbook of Global Indigenous Archaeology*. Oxford University Press, Oxford.

Milano, S., S. Lindauer, A. Prendergast, E. Hill, C. Hunt, G. Barker and B. Schöne 2018. Mollusk carbonate thermal behaviour and its implications in understanding prehistoric fire events in shell middens. *Journal of Archaeological Science: Reports* 20: 443–457.

Milano, S., A.L. Prendergast and B.R. Schöne 2016. Effects of cooking on mollusk shell structure and chemistry: Implications for archeology and paleoenvironmental reconstruction. *Journal of Archaeological Science Reports* 7: 14–26.

Mindeguia, J. C., C. Sirieix, A. Brodard, P. Guibert, C. Ferrier, P. Malaurent and D. Lacanette 2013. Simulation de l'impact thermomécanique d'un feu en carrière calcaire souterraine: Implications pour l'étude des feux paléolithiques de la grotte Chauvet-Pont-d'Arc (Ardèche). *21ème Congrès Français de Mécanique (Bordeaux, 2013)*. Courbevoie: Association Française de Mécanique. https://hal.archives-ouvertes.fr/hal-03441390/document.

Mobley, C.M. and M. Eldridge 1992. Culturally modified trees in the Pacific Northwest. *Arctic Anthropology* 29(2): 91–110.

Moody, D. 1976. Thermal Alteration of Quartzite from Spanish Diggings, Wyoming - A Pre-Historic Quarry. *Transactions of the Nebraska Academy of Sciences and Affiliated Societies* 409: 8–11.

Morgan, P. 2013. *The Settling of Gippsland: A Regional History*. Traralgon: Gippsland Local Government Network.

Morrison, M. and E. Shepard 2013. The archaeology of culturally modified trees: Indigenous economic diversification within colonial intercultural settings in Cape York Peninsula, northeastern Australia. *Journal of Field Archaeology* 38(2): 143–160.

Mphuthi, D.D., R. Pretorius and C. Van Der Walt 2013. Coping with haemodialysis: a mixed method study of families in Gauteng. *Journal of Renal Nursing* 5(3): 134–140.

Mulligan, J. 2020. Fire in East Gippsland: Recollections of John Mulligan. *Volunteer Fire Fighters Association* online website. https://volunteerfirefighters.org.au/?s=john+mulligan and https://volunteerfirefighters.org.au/fire-in-east-gippsland-recollections-of-john-mulligan.

Murphy, B.P., R.A. Bradstock, M.M. Boer, J. Carter, G.J. Cary, M.A. Cochrane, R.J. Fensham, J. Russell-Smith, G.J. Williamson, D.M.J.S. Bowman 2013. Fire regimes of Australia: A pyrogeographic model system. *Journal of Biogeography* 40(6): 1048–1058.

Navin Officer Heritage Consultants Pty Ltd 2015. Moorebank Intermodal Terminal: Aboriginal Heritage Assessment—Addendum: Scarred Tree Assessment (MA6 and MA7). Unpublished report, Navin Officer Heritage Consultants Pty Ltd, Kingston.

Neubauer, F. 2018. Use-alteration analysis of fire-cracked rocks. *American Antiquity* 83(4): 681–700.

Nicholson, R.A. 1993. A morphological investigation of burnt animal bone and an evaluation of its utility in archaeology. *Journal of Archaeological Science* 20: 411–428.

O'Connor S., J. Balme, J. Fyfe, J. Oscar, M. Oscar, J. Davis, H. Malo, R. Nuggett and D. Surprise 2013. Marking resistance? Change and continuity in the recent rock art of the southern Kimberley, Australia. *Antiquity* 87: 539–54.

Oertle, A. 2019. Time and relative dimension in space: Untangling site formation and taphonomic processes on archaeological shell from the tropical Indo-Pacific. Unpublished PhD dissertation, University of Sydney.

Olson, J.S. 1963. Energy storage and the balance of producers and decomposers in ecological systems. *Ecology* 44(2): 322–331.

Opuchovic, O. and A. Kareiva 2015. Historical hematite pigment: Synthesis by an aqueous sol–gel method, characterization and application for the colouration of ceramic glazes. *Ceramics International* 41(3): 4504–4513.

Östlund, L., L. Ahlberg, O. Zackrisson, I. Bergman and S. Arno 2009. Bark-peeling, food stress and tree spirits—The use of pine inner bark for food in Scandinavia and North America. *Journal of Ethnobiology* 29(1): 94–112.

Östlund, L., B. Keane, S. Arno and R. Andersson 2005. Culturally scarred trees in the Bob Marshall wilderness, Montana, USA—Interpreting native American historical forest use in a wilderness area. *Natural Areas Journal* 25(4): 315–325.

Padoch, C., E. Harwell and A. Susanto 1998. Swidden, Sawah, and In-Between: Agricultural Transformation in Borneo. *Human Ecology* 26: 4–20.

Pagoulatos P. 2005. Experimental Burned Rock Studies on the Edwards Plateau: A View from Camp Bullis, Texas. *North American Archaeologist* 26(3): 289–329.

Pascoe, B. 2021. *Dark Emu: Aboriginal Australia and the Birth of Agriculture*. Broome: Magabala Books.

Passe-Courtin, N., P. N´Guyen, R. Pemard, A. Ouensanga and C. Bouchon 1995. Water desorption and aragonite-calcite phase transition in scleractinian coral skeletons. *Thermochimica Acta* 265: 135–140.

Patterson, L.W. 1995. Thermal Damage of Chert. *Lithic Technology* 20(1): 72–80.

Pausas, J.G. *and J.E.* Keeley 2009. A burning story: the role of fire in the history of life. *BioScience* 59*:* 593–601.

Penman, T. D., D. Binns, R. Allen, R. Shiels and S. Plummer 2008. Germination responses of a dry sclerophyll forest soil-stored seedbank to fire related cues. *Cunninghamia: a journal of plant ecology for eastern Australia* 10(4): 547–555.

Pepper, P. 1980. *You are what you make yourself to be. The story of a Victorian Aboriginal family 1842-1980*. Melbourne: Hyland House.

Pétrequin, P. and A.M. Pétrequin 2002. *Écologie d'un outil : la hache de pierre en Irian Jaya (Indonésie)*. Nouvelle édition, First published 1993, Monographie du CRA12, CNRS Éditions.

Picha, P.R., S.A. Ahler, R.D. Sayler and R.W. Seabloom. 1991. Effects of prairie fire on selected artifact classes. *Archaeology in Montana* 32(2): 15-28.

Plomley, N.J.B. 1966. *Friendly mission: The Tasmanian historical papers of George Augustus Robinson 1829-1834*. Hobart: Tasmanian Historical Research Association.

Pomiès, M.P., M. Menu and C. Vignaud 1999. Red Palaeolithic pigments: natural hematite or heated goethite? *Archaeometry* 41(2): 275–285.

Porter, L. 2006. Rights or containment? The politics of Aboriginal cultural heritage in Victoria. *Australian Geographer* 37(3): 355–374.

Price, O.F. and R.A. Bradstock 2013. The spatial domain of wildfire risk and response in the Wildland Urban Interface in Sydney, Australia. *Natural Hazards and Earth System Sciences* 13(12): 3385–3393.

Pullin, R. 2011. *Eugene von Guérard: Nature Revealed*. Melbourne: National Gallery of Victoria.

Pullin, R. 2018. *The Artist as Traveller: The Sketchbooks of Eugene von Guérard*: 214–221. Ballarat: Art Gallery of Ballarat.

Purdy, B.A. 1971. Investigations Concerning the Thermal Alteration of Silica Minerals: An Archaeological Approach. Unpublished PhD dissertation, University of Florida.

Pyne, S.J. 1991. *Burning bush: A fire history of Australia*. New York: Henry Holt and Company.

Raper, E. 2003. 'Up in Flames': A Benchmark Investigation of the Long-Term Impacts of Fire on Rock-Art and the Development of a Rock-Art Fire Vulnerability Index. Unpublished BA(Hons) dissertation, University of Sydney.

Rebbeck, J., D. Yaussy, L. Iverson, T. Hutchinson, R. Long, A. Bova, and M. Dickinson 2006 Use of temperature-sensitive paints as an index of heat output from fire, in M.B. Dickinson (ed.) *Fire in eastern oak forests: Delivering science to land managers. Proceedings of a conference, 2005 November 15-17, Columbus, OH. Gen. Tech. Rep. NRS-P-1.* Newtown Square, PA: US Department of Agriculture, Forest Service, Northern Research Station: 295.

Reynolds, T., G. Barker, H. Barton, G. Cranbrook, L. Farr, C. Hunt, L. Kealhofer, V. Paz, A. Pike, P.J. Piper, R.J. Rabett, G. Rushworth, C. Stimpson and K. Szabó 2013. The First Modern Humans at Niah, *c.* 50,000–35,000 Years Ago, in G. Barker (ed.) *Rainforest foraging and farming in Island Southeast Asia, The Archaeology of the Niah Caves* Volume 1: 135–172. Cambridge: McDonald Institute for Archaeological Research.

Rhoads, J.W. 1992. Significant sites and non-site archaeology: A case-study from south-east Australia. *World archaeology* 24(2): 198–217.

Ribber, S. 1940. A 'boob' in a baob tree: A queer lock-up. *Sydney Morning Herald* 31 August, p. 9.

Rifkin, R.F. 2011. Assessing the efficacy of red ochre as a prehistoric hide tanning ingredient. *Journal of African Archaeology* 9(2): 131–158.

Robinson, G.A., G. Mackaness and G.H. Haydon 1941. *George Augustus Robinson's journey into south eastern Australia: With George Henry Haydon's narrative of part of the same journey*. Sydney: G. Mackaness.

Robinson, J. R. and J. D. Kingston. 2020. Burned by the fire: Isotopic effects of experimental combustion of faunal tooth enamel. *Journal of Archaeological Science: Reports* 34: 102593. DOI: 10.1016/j.jasrep.2020.102593.

Rockman, M., M. Morgan, S. Ziaja, G. Hambrecht and A. Meadow 2016. Cultural Resources Climate Change Strategy. Washington, DC: Cultural Resources, Partnerships, and Science and Climate Change Response Program, National Park Service.

Royal Society of Victoria 1891. Minutes from the meeting of 10 July 1891. *Proceedings of the Royal Society of Victoria* 3(new series): 124-129.

Rude, T. and A. Trinkle Jones. 2012. Chapter 3: Fire Effects on Prehistoric Ceramics, in K.C. Ryan, A. Trinkle, C.L. Koerner, K.M. Lee (eds) *Wildland Fire in Ecosystems: Effects of Fire on Cultural Resources and Archaeology*: 85-96. General Technical Report RMRS-GTR 42, volume 3. Fort Collins (CO): US Department of Agriculture, Forest Service, Rocky Mountain Research Station.

Russell-Smith, J., P.J. Stanton, P.J. Whitehead and A. Edwards 2004. Rain forest invasion of eucalypt-dominated woodland savanna, Iron Range, north-eastern Australia: I. Successional processes. *Journal of Biogeography*, 31: 1293–1303.

Ryan, K.C., C.L. Koerner, K.M. Lee and N. Siefkin 2012. Chapter 1: Effects of Fire on Cultural Resources—Introduction, in K.C. Ryan, A. Trinkle, C.L. Koerner, K.M. Lee (eds) *Wildland Fire in Ecosystems: Effects of Fire on Cultural Resources and Archaeology*: 1–14. General Technical Report RMRS-GTR 42, volume 3. Fort Collins (CO): US Department of Agriculture, Forest Service, Rocky Mountain Research Station.

Ryan, K.C., A. Trinkle Jones, C.L. Koerner and K.M. Lee (eds) 2012. *Wildland fire in ecosystems: effects of fire on cultural resources and archaeology*. General Technical Report RMRS-GTR-42 volume 3. Fort Collins (CO): U.S. Department of Agriculture, Forest Service, Rocky Mountain Research Station.

Sagona, A.G. 1994. The quest for red gold, in A. G. Sagona (ed.) *Bruising the red earth: Ochre mining and ritual in Aboriginal Tasmania*: 8–38. Carlton: Melbourne University Press.

Salmon, F. 2019. Simulation aéro-thermo-mécanique des effets du feu sur les parois d'un milieu confiné: Application à l'étude des thermo-altérations de la grotte Chauvet-Pont d'Arc. Unpublished PhD dissertation, Université de Bordeaux.

Salomon, H., C. Vignaud, Y. Coquinot, L. Beck, C. Stringer, D. Strivay and F. D'Errico 2012. Selection and heating of colouring materials in the Mousterian level of Es-Skhul (*c.* 100 000 years B.P., Mount Carmel, Israel). *Archaeometry* 54(4): 698–722.

Salomon, H., C. Vignaud, S. Lahlil and N. Menguy 2015. Solutrean and Magdalenian ferruginous rocks heat-treatment: Accidental and/or deliberate action? *Journal of Archaeological Science* 55: 100–112.

Santana, V.M., R.A. Bradstock, M.K. Ooi, A.J. Denham, T.D. Auld and M.J. Baeza 2010. Effects of soil temperature regimes after fire on seed dormancy and germination in six Australian Fabaceae species. *Australian Journal of Botany* 58(7): 539–545.

Scadding, R., V. Winton and V. Brown 2015. An LA-ICP-MS trace element classification of ochres in the Weld Range environ, Mid West region, Western Australia. *Journal of Archaeological Science* 54: 300–312.

Schmidt P. 2019. How reliable is the visual identification of heat treatment on silcrete? A quantitative verification with a new method. *Archaeological and Anthropological Sciences* 11: 713–726.

Schmidt, P., L. Bellot-Gurlet and H. Floss 2018. The unique Solutrean laurel-leaf points of Volgu: heat-treated or not? *Antiquity* 92(363): 587–602.

Schmidt, P., Buck, G., Berthold, C., Lauer, C. and K. Nickel 2019. The Mechanical Properties of Heat-Treated Rocks: A Comparison between Chert and Silcrete. *Archaeological and Anthropological Sciences* 11(6): 2489–2506.

Schmidt, P. and P. Hiscock 2019. Evolution of Silcrete Heat Treatment in Australia—a Regional Pattern on the South-East Coast and Its Evolution over the Last 25 ka. *Journal of Paleolithic Archaeology* 2: 74–97.

Schmidt, P. and P. Hiscock 2020. Early silcrete heat treatment in Central Australia: Puritjarra and Kulpi Mara. *Archaeological and Anthropological Sciences* 12, 188: 1–7.

Schmidt, P., S. Masse, G. Laurent, A. Slodczyk, E. Le Bourhis, C. Perrenoud, J. Livage and F. Fröhlich 2012. Crystallographic and Structural Transformations of Sedimentary Chalcedony and Flint upon Heat treatment. *Journal of Archaeological Science* 39(1): 135–144.

Schmidt, P., Nash, D.J., Coulson, S., Goden, M.B. and G.J. Awcock 2017. Heat treatment as a universal technical solution for silcrete use? A comparison between silcrete from the

Western Cape (South Africa) and the Kalahari (Botswana). *PLoS ONE* 12(7): e0181586. DOI: 10.1371/journal.pone.0181586.

Schmidt, P., C. Paris and L. Bellot-Gurlet 2016. The investment in time needed for heat treatment of flint and chert. *Archaeological and Anthropological Sciences* 8: 839–848.

Schmidt, P., O. Spinelli Sanchez and C.J. Kind 2017. Stone Heat Treatment in the Early Mesolithic of Southwestern Germany: Interpretation and Identification. *PLoS ONE* 12(12): e0188576. DOI: 10.1371/journal.pone.0188576.

Sefton, C. 2011. The effects of fire on the rock art of the Woronora Plateau. *Rock Art Research* 28(1): 49–52.

Shakesby, R.A., P.J. Wallbrink, S.H. Doerr, P.M. English, C.J. Chafer, G.S. Humphreys, W.H. Blake and K.M. Tomkins 2007. Distinctiveness of wildfire effects on soil erosion in south-east Australian eucalypt forests assessed in a global context. *Forest Ecology and Management* 238(1-3): 347–364.

Sharples, J.J., R.H. McRae, R.O. Weber and A.M. Gill 2009. A simple index for assessing fire danger rating. *Environmental Modelling and Software* 24(6): 764–774.

Shipman, P., G. Foster and M. Schoeninger 1984. Burnt bones and teeth: An experimental study of color, morphology, crystal structure and shrinkage. *Journal of Archaeological Science* 11: 307–325.

Shvidenko, A.Z. and D.G. Schepaschenko 2013. Climate Change and Wildfires in Russia. *Contemporary Problems of Ecology* 6: 683–692.

Sillitoe, P. 1988. *Made in Niugini: Technology in the Highlands of Papua New Guinea*. London: British Museum Publications.

Skelly, R., J. Crouch, B. David, R. Mullett and J. Fresløv 2019. Holey Plains State Park and Nunnett-Timbarra River area: Post-bushfire cultural heritage surveys, 9 July-14 August 2019. Unpublished report to the GunaiKurnai Land and Waters Aboriginal Corporation, Kalimna West.

Skene, A.J. and R.B. Smyth 1874. *Report on the physical character and resources of Gippsland*. Melbourne: John Ferres.

Smith, J. 1876. *Argus* 24 July: 5, column 4.

Smith, L. 2000. A history of Aboriginal heritage legislation in south-eastern Australia. *Australian Archaeology* 20(50): 109–118.

Smith, M.A., B. Fankhauser and M. Jercher 1998. The changing provenance of red ochre at Puritjarra rock shelter, central Australia: Late Pleistocene to present. *Proceedings of the Prehistoric Society* 64: 275–292.

Smyth, R.B. 1878. *Aborigines of Victoria, with Notes Relating to the Habits of the Natives of Other Parts of Australia and Tasmania Compiled from Various Sources for the Government of Victoria*. Melbourne: John Ferres, Government Printer.

Solari, A., D. Olivera, I. Gordillo, P. Bosch, G. Fetter, V.H. Lara and O. Novelo 2015. Cooked bones? Method and practice for identifying bones treated at low temperature. *International Journal of Osteoarchaeology* 25: 426–440.

Speiss, A.E. 1988. On New England shell middens: Response to Sanger's cautionary tale. *American Antiquity* 53: 174–177.

Spry, C. 2016. Evaluating the recording system for high- and low-density stone artefact occurrences in Victoria, Australia: A stone artefact analysis perspective. *Excavations, Surveys and Heritage Management in Victoria* 4: 43–51.

Spurrell, W. G. 1976? *History of Nicholson, 1876-1976*. Place of publication and publisher unknown.

Stahl, P.W. 1996. The recovery and interpretation of microvertebrate bone assemblages from archaeological contexts. *Journal of Archaeological Method and Theory* 3: 31–75.

Stiner, M.C. 1991. Food procurement and transport by human and non-human predators. *Journal of Archaeological Science* 18: 455–482.

Stiner, M.C., S.L. Kuhn, S. Weiner and O. Bar Yosef 1995. Differential burning, recrystallization, and fragmentation of archaeological bone. *Journal of Archaeological Science* 22: 223–237.

Stout, D. 2002. Skill and Cognition in Stone Tool Production. *Current Anthropology* 43(5): 693–722.

Straw, E.E. 1956. The Hazeldeane Selection. In the Burchett papers, MS 7997, State Library of Victoria, Melbourne.

Stretton, L. 1939. *Report of the Royal Commission to Inquire into the Causes of and Measures Taken to Prevent the Bush Fires of January, 1939 and to Protect Life and Property and the Measures to be Taken to Prevent Bush Fires in Victoria and to Protect Life and Property in the Event of Future Bush Fires*. Melbourne: Victorian Government.

de Strzelecki, P.E. 1841. Report by Count Streleski, in 'Copy of a despatch from Sir G. Gipps, Governor of New South Wales, to the Secretary of State for the colonies, transmitting a report of the progressive discovery and occupation of that colony during the period of his administration of the Government. House of Commons Papers 17(1): 11–17.

de Strzelecki, P.E. 1845. *Physical description of New South Wales and Van Diemen's Land: Accompanied by a geological map, sections and diagrams, and figures of the organic remains*. London: Longman, Brown, Green, and Longmans.

Sumner, P.D., K.J. Hall, J.L. van Rooy and K.I. Meiklejohn 2009. Rock weathering on the eastern mountains of southern Africa: Review and insights from case studies. *Journal of African Earth Sciences* 55: 236–244.

Swete Kelly, M.C. and S. Phear 2004. Report on the salvage of materials from the Weston Storage Facility, Department of Archaeology and Natural History, Australian National University. Unpublished report filed at https://www.museum-sos.org/docs/WestonStoreReportPart1.pdf.

Symes, S.A., C.W. Rainwater, E.N. Chapman, D.R. Gipson and A.L. Piper 2008. Patterned thermal destruction of human remains in a forensic setting, in C.W. Schmidt and S.A. Symes (eds) *The Analysis of Burned Human Remains*: 15–54. Burlington (MA): Academic Press.

Szabó, K. 2005. Technique and Practice: Shell Working in the Western Pacific and Island Southeast Asia. Unpublished PhD dissertation, Australian National University.

Szabó, K. 2017. Molluscan shells as raw materials for artefact production, in M. Allen (ed.) *Molluscs in Archaeology: Methods, approaches and applications*: 308–325. Oxford: Oxbow Books.

Szabó, K. and B. Koppel. No date. Unpublished data from a shell-working workshop, 2012. Armidale NSW, Australia.

Taçon, P.S.C. and D. Hardin 2019. The aftermath of fire damage to important rock art at Baloon Cave, Carnarvon Gorge, Queensland. Unpublished paper presented at the Australian Archaeological Association annual conference, Gold Coast subsequently posted on https://www.bradshawfoundation.com/rockartnetwork/baloon_cave.php.

Tanjil 1886. *The centennial guide to the Gippsland lakes and rivers*. Melbourne: M. L. Hutchinson.

The Mountain Cattlemen's Association of Victoria 2015. The links between cattle grazing and fuel reduction in the grazing zones of the High Country. February. Unpublished report by The Mountain Cattlemen's Association of Victoria. https://www.mcav.com.au/assets/files/news/20150327-the-links-grazing-and-fuel-2010-and-2015.pdf.

Thompson, D.M. 1952. Forest fires prevention and control in the Cann Valley forest district. Unpublished Diploma thesis, University of Melbourne.

Thompson, K. 1985 *A history of the Aboriginal people of East Gippsland.* Melbourne: Land Conservation Council.

Thompson, T.J.U. 2004. Recent advances in the study of burned bone and their implications for forensic anthropology. *Forensic Science International* 1465: S203–S205.

Thomson, D. 2005 *Donald Thomson in Arnhem Land.* Compiled and introduced by Nicolas Peterson. Carlton: The Miegunyah Press, Melbourne University Publishing.

Tibbett, K. 2005. Community specialisation, standardisation and exchange in a hunter-gatherer society: A case study from Kalkadoon country, northwest Queensland, Australia. Unpublished PhD dissertation, James Cook University.

Timmons, R.S., L. deBano and K.C. Ryan 2012. Chapter 9: Implications of Fire Management on Cultural Resources, in K.C. Ryan, A. Trinkle Jones, C.L. Koerner and K.M. Lee (eds) *Wildland fire in ecosystems: effects of fire on cultural resources and archaeology*: 171–191. General Technical Report RMRS-GTR-42 volume 3. Fort Collins (CO): U.S. Department of Agriculture, Forest Service, Rocky Mountain Research Station.

Toffolo, M.B. 2021. The significance of aragonite in the interpretation of the microscopic archaeological record. *Geoarchaeology: An International Journal* 36: 149–169.

Tonkin, D. and C. Landon 1999. *Jackson's Track: Memoir of a Dreamtime place.* Melbourne: Penguin.

Tratebas, A.M., N.V. Cerveny and R.I. Dorn 2004. The effects of fire on rock art: Microscopic evidence reveals the importance of weathering rinds. *Physical Geography* 25(4): 313–333.

Tutchener, D. and R. Kurpiel 2021. Proximity of Aboriginal Cultural Heritage Places to fresh and salt water in the Bunurong Land Council Aboriginal Corporation Registered Aboriginal Party area: preliminary GIS analysis. *Excavations, Surveys and Heritage Management in Victoria* 10: 47–51.

Twidale, C.R. 1968. *Geomorphology: with special reference to Australia.* Melbourne: Nelson.

United States National Park Service 2014. *Climate Change and Stewardship of Cultural Resources.* US DOI National Park Service Policy Memorandum 14-02. Electronic document http://www.nps.gov/policy/PolMemos/PM-14-02.htm, accessed 29 August 2022.

Vanderwal, R. (ed.) 1994. *John Bulmer's Recollections of Victorian Aboriginal Life, 1855-1908.* Melbourne: Museum of Victoria.

Velo, J. 1984. Ochre as medicine: a suggestion for the interpretation of the archaeological record. *Current Anthropology* 25(5): 674–674.

Victorian Traditional Owner Cultural Fire Knowledge Group 2019. *The Victorian Traditional Owner Cultural Fire Strategy.* https://www.ffm.vic.gov.au/bushfire-fuel-and-risk-management/traditional-owner-burns, accessed 4 September 2022.

Villagran, X.S. 2014. Experimental micromorphology on burnt shells of *Anomalocardia brasiliana* (Gmelin 1791) (Bivalvia, Veneridae) and its potential for identification of combustion features on shell-matrix sites. *Geoarchaeology: An International Journal* 29: 389–396.

Villagran, X.S. and R.M. Poch 2014. A new form of needle-fiber calcite produced by physical weathering of shell. *Geoderma* 213: 173–177.

Vinnicombe, P. 1997. Aspects of the value system associated with Wandjina art, in K.F. Kenneally, M.R. Lewis, M. Donaldson, and C. Clement (eds.) *Aboriginal rock art of the Kimberley: Proceedings of a seminar held at the University of Western Australia, Perth, 8 March 1997*: 13–18. Occasional Paper No. 1. Perth: Kimberley Society.

Wadley, L. 2009. Post-depositional heating may cause over-representation of red-coloured ochre in Stone Age sites. *South African Archaeological Bulletin* 64(190): 166–171.

Waikato Radiocarbon Dating Laboratory. 2017. AMS Processing Technical Report. URL: https://radiocarbondating.com/__data/assets/pdf_file/0010/387712/Waikato-Radiocarbon-Dating-Laboratory-AMS-Processing-Technical-Report-2017.pdf Accessed 6.2.2022.

Wallis, L.A., D.A. Argue and M. Pearson 2003. Heritage and bushfires in the ACT. *Heritage in Trust* Winter 2003: 5–8.

Wallis, L., J. Huntley, M. Marsh, A. Watchman, A. Ewen and A. Strano 2016. PXRF analysis of a yellow ochre quarry and rock art motifs in the Central Pilbara. *Journal of the Anthropological Society of South Australia* 40: 134–155.

Wakefield, N.A. 1970. Bushfire frequency and vegetational change in south-eastern Australian forests. *Victorian Naturalist* 87: 152–158.

Walsh, M.K., H.J. Duke and K.C. Haydon 2018. Toward a better understanding of climate and human impacts on late Holocene fire regimes in the Pacific Northwest, USA. *Progress in Physical Geography: Earth and Environment* 42: 478–512.

Wardlaw, J. 1997. Alfred William Howitt: Scientific pioneer in colonial Victoria. *Victorian Historical Journal* 248(68): 79–93.

Watchman, A. 1993. Perspectives and potentials for absolute dating prehistoric rock paintings. *Antiquity* 67(254): 58–65.

Watchman, A. 2004a. Short-Term Impacts of Water and Frost on the Fire-Affected Rendezvous Creek and Nursery Swamp 2 Aboriginal Art Sites in Namadgi National Park. Unpublished report prepared for Environment ACT.

Watchman, A. 2004b. Impacts of Water and Frost on the Fire-Affected Rendezvous Creek Aboriginal Art Site in Namadgi National Park. Unpublished report prepared for Environment ACT.

Waters, D.A., G.E. Burrows and J.D. Harper 2010. *Eucalyptus regnans* (Myrtaceae): A fire-sensitive eucalypt with a resprouter epicormic structure. *American Journal of Botany* 97(4): 545–556.

Watson, D. 1984. *Caledonia Australis: Scottish highlanders on the frontier of Australia*. Sydney: Collins.

Watson, D. 2016 *The Bush*. Melbourne: Penguin.

Webster, C. 2021. Mineralogical and geochemical analysis of Cloggs Cave rock art. Unpublished BA(Hons) thesis, Griffith University.

Westerling, A.L., H.G. Hidalgo, D.R. Cayan and T.W. Swetnam 2006. Warming and Earlier Spring Increase Western U.S. Forest Wildfire Activity. *Science* 313: 940–943.

Westgarth, W. 1848. *Australia Felix: Or, a historical and descriptive account of the settlement of Port Phillip, New South Wales: Including full particulars of the manners and condition of the aboriginal natives: With observations on emigration, on the system of transportation, and on colonial policy*. Edinburgh: Oliver and Boyd.

Wotton, B.M., J.S. Gould, W.L. McCaw, N.P. Cheney and S.W. Taylor 2011. Flame temperature and residence time of fires in dry eucalypt forest. *International Journal of Wildland Fire* **21**: 270–281.

Zazzo, A., J.-F Saliége, M. Lebon, S. Lepetz and C. Moreau 2012. Radiocarbon dating of calcined bones: insights from combustion experiments under natural conditions. *Radiocarbon* 54: 855–866.

Zhang, W., Q. Sun, S. Zhu and B. Wang 2017. Experimental study on mechanical and porous characteristics of limestone affected by high temperature. *Applied Thermal Engineering* 110: 356–62.